B

Experientia Supplementum
Vol. 44

Birkhäuser Verlag
Basel · Boston · Stuttgart

Nutritional Adequacy, Nutrient Availability and Needs

Nestlé Nutrition Research Symposium,
Vevey, September 14–15, 1982

Edited by
J. Mauron

1983

Birkhäuser Verlag
Basel · Boston · Stuttgart

The symposium was organized and sponsored by the
Research Department of Nestlé Products Technical Assistance Co. Ltd.,
CH-1814 La Tour-de-Peilz (Switzerland)

Volume Editor
Prof. J. Mauron
Research Department
Nestlé Products Technical Assistance Co. Ltd.
P. O. Box 88
CH-1814 La Tour-de-Peilz (Switzerland)

Editorial Board
K. Anantharaman
P. A. Finot
M. Horisberger
Y. Ingenbleek
H. P. Würzner

CIP-Kurztitelaufnahme der Deutschen Bibliothek

Nutritional adequacy, nutrient availability and needs :
Nestlé Nutrition research symposium, Vevey, September
14—15, 1982 / ed. by J. Mauron. [The syposium was
organized and sponsored by the Research Dep. of Nestlé
Products Techn. Assistance Co. Ltd., La Tour-de-Peilz
(Switzerland). Ed. board K. Anantharaman ...]. –
Basel ; Boston ; Stuttgart : Birkhäuser. – 1983.
    (Experientia : Suppl. ; Vol. 44)
    ISBN 3-7643-1479-6

NE: Mauron, Jean [Hrsg.]; Nestlé Nutrition SA ‹Vevey› ;
Experienta / Supplementum

© 1983 Birkhäuser Verlag Basel
Printed in Switzerland
ISBN 3-7643-1479-6

C O N T E N T S

Introduction
    J. Mauron   7

How much food does man require ? An evaluation of human
energy needs
    E.M. Widdowson   11

Thermogenic responses induced by nutrients in man :
Their importance in energy balance regulation
    E. Jéquier   26

Energy expenditure and whole body protein synthesis
in very low birth weight (VLBW) infants
    Y. Schutz, C. Catzeflis, F. Gudinchet, J. Micheli,
    C. Welsch, M.J. Arnaud and E. Jéquier   45

Energy fuel and hormonal profile in experimental obesities
    B. Jeanrenaud   57

Nutrient intake and energy regulation in physical exercise
    H. Howald and J. Décombaz   77

Protein turnover, nitrogen balance and rehabilitation
    E.B. Fern and J.C. Waterlow   89

Amino acid signals and food intake and preference : Relation
to body protein metabolism
    A.E. Harper and J.C. Peters   107

Food processing and storage as a determinant of protein
and amino acid availability
    R.F. Hurrell and P.A. Finot   135

6   Contents

Energy/protein interrelation in experimental food restriction
    K. Anantharaman   157

Behavioural strategies in the regulation of food choice
    P.D. Leathwood and D.V.M. Ashley   171

The relationship of pellagra to corn and the low availability
of niacin in cereals
    K.J. Carpenter   197

Iron requirements and bioavailability of dietary iron
    L. Hallberg   223

Vitamin deficiencies in rice-eating populations.
Effects of B-vitamin supplements
    M.S. Bamji   245

Vitamin A-deficiency impairs the normal mannosylation,
conformation and iodination of thyroglobulin : A new
etiological approach to endemic goitre
    Y. Ingenbleek   264

Is the adult protein-energy malnutrition syndrome the same
as that described in the infant ?
    J. Mauron and I. Antener   298

Public health/clinical significance of inorganic chemical
elements
    M. Abdulla   339

Food consumption, neurotransmitter synthesis, and
human behaviour
    R.J. Wurtman   356

General remarks - Some personal reflections
    A.E. Bender   370

I N T R O D U C T I O N

The establishment of the nutritional requirements of man in qualitative and quantitative terms has been the preoccupation of nutritionists since the emergence of nutrition as a science starting with Lavoisier. The accumulation of knowledge during almost two centuries led to the definition of nutritional adequacy in terms of so-called "recommended dietary allowances" established and periodically revised by committees of experts in the field.

These recommended dietary allowances are of a great help to the nutritionists who have to evaluate the nutritional adequacy of the food supply of populations or groups of people but do not represent a final answer as regards the nutrient needs of the individual in his particular living conditions. It is somewhat unfortunate that the apparent precision of the recommended dietary allowances has led many people, even nutritionists, to believe that we know today what nutritional adequacy means and that only a few secondary problems remain to be solved. Well, if this was the case, we would not have to organize symposia like the present one.

The fact is that in spite of the false security provided by the recommended allowances, our knowledge about the true nutritional requirements of men is still rather unprecise.

The present symposium gives a good picture of some of the difficulties encountered when trying to define nutritional adequacy and nutrient needs.

Nutritional individuality is a characteristic of man for which there is now increasing experimental evidence. It is a direct consequence of the biochemical individuality of each of us. Several papers touch upon this subject in a direct or indirect way : the evaluation of human energy needs; the thermogenic responses induced by nutrients in man; assays with animal models relating experimental obesity to hormonal profile and studies on metabolic and behavioural mechanisms regulating individual food choice in rats.

Another important factor of uncertainty in judging nutritional adequacy is the phenomenon of adaptation; to high or low nutrient supply, to environmental stress, to increased energy expenditure, etc.. The considerable versatility for physiological adaptation renders the actual definition of "true" needs very difficult. An example of physiological adaptation to increased energy expenditure is given in the paper on nutrient intake and energy regulation in physical exercise, whereas the breakdown of adaptation is shown in the description of the adult protein energy-malnutrition syndrome. As to the mechanism of recovery from malnutrition, a model is presented on protein turnover and nitrogen balance during rehabilitation in human and animal protein deficiency. The role of environmental stress on vitamin requirements is shown in a paper on vitamin deficiencies in rice-eating populations.

The metabolic interrelation between nutrients also influences nutrient needs. Two papers deal with energy-protein interdependence : one during the rapid growing period of low birth weight infants, the other in rats during experimental food restriction. Another paper establishes a new, fundamental interrelation between nutrients, namely the link between iodine- and vitamin A-deficiencies in the aetiology of endemic goitre.

A crucial aspect of nutritional adequacy is the actual biological availability of nutrients. Indeed, many studies on nutrient needs are based on nutrient content of the food that does not always correspond to actual biological availability. Several presentations deal with the subject. One refers to the low availability of niacin in cereals, the other discusses the bioavailability of dietary iron as it is influenced by meal composition and the third paper deals with the effect of food processing on amino acid availability.

At last, new frontiers relating to nutritional adequacy are tackled. One subject is the biological role of many trace elements that is still a matter of debate among nutritionists. The other concerns the interrelation between certain nutrients and brain function and its feedback on behaviour and food intake. One paper elaborates on the mechanisms involved in reducing energy intake when an unbalanced protein is fed, the other discusses the impact certain nutrients may have on behaviour, as precursors of neurotransmitters. This kind of research introduces a new point of view, considering nutritional adequacy not only in respect to fulfilling basic nutrient needs for growth and maintenance but also as regards optimal functioning of the organism and subjective well-being.

I would like to thank here Dr. K. Anantharaman, the initiator of the Symposium, who in spite of my initial reluctance, pushed the idea forward and obtained the agreement of the Management.

I am grateful to Dr. C.L. Angst, General Manager of Nestlé, who rendered the Symposium possible by generous financial support.

Many people contributed to the success of this Symposium and they cannot all be mentioned here. I convey my special thanks to all the persons who worked somehow behind the scene but were indispensable for the final success:

Mrs J. Vocanson for the initial organisation and convocation,
Miss C. Mordasini for the typing of the manuscripts,
Mrs J. Farr for polishing the English of the presentations,
Mrs M. Beaud and Mr. K. Fleury for the drawing and retouching of the figures
and Mrs J. Jaquier for controlling and completing the references.

Finally, I would like to thank the editorial board and all the scientists who made the Symposium such a success by their excellent presentations.

Professor Jean Mauron
Head, Nestlé Research Department
Editor-in-Chief

HOW MUCH FOOD DOES MAN REQUIRE ?
AN EVALUATION OF HUMAN ENERGY NEEDS

Elsie M. WIDDOWSON

University of Cambridge Clinical School, Department of Medicine,
Level 5, Addenbrooke's Hospital, Cambridge, England

SUMMARY

Nutritional individuality is a characteristic of mankind and this is as true of energy intakes and needs as of other attributes. Studies over the years have shown that individuals vary by a factor of two or more in their intakes of energy from the first year after birth to 75 years and over. The metabolic differences that must lie behind this are still not fully understood. Recent ideas about the importance of dietary thermogenesis in energy expenditure seemed as though they might provide an explanation, but not all investigators agree, and the problem has still not been satisfactorily solved.

In spite of these individual variations, average intakes and expenditures show predictable changes with age, and males always seem to take more energy than females throughout their lives. This is still true when the intakes are expressed per kg body weight, per square metre surface area and per kg lean body mass. The explanation of this is not certain, but various suggestions have been made.

There is evidence that energy intakes, at any rate in the
United Kingdom, were falling between the 1950's and 1970's, and
this has been attributed to a decrease in physical activity.
However, some of the apparent difference is due to the fact that
higher factors were used to calculate the energy derived from
protein, fat and carbohydrate in the 1930's and 1940's than were
used later; if the later factors are applied to the results of
the earlier surveys the values for energy are reduced by about
10%. This correction brings the results of the earlier surveys
into line with those of the later ones for boys up to 14 years
and girls up to 10 years. Older children of both sexes and
adults, however, do seem to be eating less than they used to do.

The question as to whether women should and do increase their
intake of food during pregnancy has been discussed at two previous meetings sponsored by Nestlé. The evidence seems to be
that they do not, and it is postulated that metabolic economies
enable the women to produce 4 kg of body fat and a foetus
weighing 3.5 kg without any increase in energy intake at all.

Infants grow very rapidly, particularly during the first
6 months after birth, and a considerable proportion of the
energy intake is directed to this end. The conventional method
of calculating energy from the intakes of protein, fat and
carbohydrate is quite inappropriate in these circumstances.

* * *

The title of my talk "How much food does man require ? An evaluation of
human energy needs" might lead you to expect some pronouncement on the
number of calories or joules people of various ages and conditions require
each day. I am not going to make any such pronouncement. Better qualified
people than I have applied themselves to this task - most recently those
contributing to the FAO/WHO Committee's Report on Energy and Protein Needs
(1982). I am going to confine myself to general principles, particularly
those on which new information is emerging, or those which present problems
that have not yet been solved.

Nutritional individuality

Nutritional individuality is a characteristic of mankind and, as with other aspects of human biology, no two individuals are the same. There may be family resemblances, but each person develops his own eating habits and, quite apart from habits, each one of us has his own individual physiological requirements for energy and for each of the various nutrients, and these vary with age, sex and external circumstances. The "average man" has never existed as an individual, but the concept of him is useful so long as we remember that the description of him represents the average of measurements made on a number of individuals, and the measurements on individuals may vary considerably between the greatest and the least.

The studies I made in the 1930's on the individual food intakes of men, women and children illustrated clearly how widely the energy intakes varied from one individual to another of the same age and sex (Widdowson, 1936, 1947; Widdowson & McCance, 1936). Again and again, two physically similar individuals might differ in their intakes of energy to the extent that the intake of one was twice that of the other. These variations were evident over the age range from one to eighteen years and on into the fifties, and more recent studies have shown that they are found in infants between birth and one year (Morgan et al., 1976; Morgan & Mumford, 1977) and in men and women aged 75 and over (Darke et al., 1980). The variation between individuals is somewhat less among groups of similar people living together, eating together and following the same programme of exercise e.g. army officer cadets in training (Edholm et al., 1955) and army recruits (Edholm et al., 1970), which suggests that differences in activity and way of life account for some of the variation. Considerable variability is, however, still evident, not only in energy intakes, but also in energy expenditure per minute at rest and at various activities (Durnin & Passmore, 1967), so that one individual expends 50% more energy than another in performing the same active or inactive task. When energy expenditure is integrated over the whole day, we still find a large range of variation. A discussion of the reasons for the variation in energy needs between individuals lies outside the scope of this paper, but they do not seem to provide an easy answer as to why some people get fat and others stay lean. Those that are fat must at some time have eaten more food than would provide for their own individual energy requirements, whether they are high or low expenders of energy.

Dietary-induced thermogenesis

This brings me to the controversial subject of dietary-induced thermogenesis. Up to a few years ago, it was believed that an adult maintained his
body weight more or less constant by a fairly exact control over his intake
of food. The question then was how appetite was regulated so accurately;
thermostatic, glucostatic, lipostatic and other mechanisms were all considered. In the last 10 years, the idea has been put forward that energy
expenditure plays an important part in maintaining a steady body weight. In
fact this idea is not new, for it has been around for 80 years (Neumann,
1902). It was revived by Miller & Mumford (1967) and given a new lease of
life by Rothwell & Stock (1979, 1982). These investigators believe that if
more energy is taken in than is required for maintenance and activity, then
some of the excess energy is lost as heat and this takes place particularly
after meals. This loss of heat is over and above the energy expenditure
involved in digesting and absorbing the nutrients in the meal, the so-called
"specific dynamic action". It has been suggested that dietary induced
thermogenesis is greatly enhanced by exercise (Miller & Wise, 1975), but
this has now been convincingly disproved (Dallosso, 1982). It was suggested
that individuals vary considerably in their ability to dissipate excess
energy in this way; those that get fat do not do it so readily as those that
stay lean (James & Trayhurn, 1981). All this implies that there is normally
some controlling heat-losing mechanism which enables the body to maintain
the status quo when the individual eats more food than he needs, but evidence for this in man has never been proved conclusively (Passmore, 1982).
Even in rats similar studies have not given the same results in the hands of
different workers. Rothwell & Stock's (1979, 1982) findings suggest that
when rats are fed what is called a cafeteria-type diet, so that they
over-eat, not all the excess energy taken in the food is laid down as fat. A
considerable part is lost as heat. However, Hervey & Tobin (1982) and
Bestley et al. (1982), apparently doing the same experiment, find that all
the excess energy taken in can be accounted for by an increase in body fat.
McCracken & Barr (1982) too have come to a similar conclusion. Until the
differences are resolved and the facts become certain it seems premature to
discuss mechanisms for dietary induced thermogenesis or to invoke brown fat
as being involved in it. Passmore (1982) has said recently "While it is

impossible to state that futile cycles, whether in BAT* or in other organs, play no part in the disposal of excess dietary energy, there is good evidence that in many people they do not operate much of the time and (when they do) only on a small scale". From all that I have read in preparing this paper, I must say, I agree with him.

In the contrary situation, when the energy intake is less than the requirement, and appetite cannot be satisfied, the only way in which the body can adapt is to reduce energy expenditure. This it does, and the metabolic rate falls in animals as well as man. The mechanism is complex and probably involves the thyroid (Jung et al., 1980). There must be variations between individuals, which will depend upon the amount by which that person's or animal's energy needs are deficient, the length of time the undernutrition has lasted and the composition of the body at the outset. One of the most striking things about this is the rapid rise in metabolic rate as soon as food is supplied.

Sex differences in energy intake

The question asked in the title of my paper "How much food does man require ?" might be regarded as rather like the question "How long is a piece of string ?". This, however, is not very helpful, and so long as we bear in mind the limitations of average figures for energy requirements, there are some things we can state with certainty. First, average intakes and expenditures, and hence requirements for energy of healthy human beings rise from birth to maturity, remain steady for a number of years and fall off gradually with advancing age. Second, the energy requirements per unit body mass fall from birth, and tend to go on doing so into old age. Third, the energy intakes of boys are higher than those of girls, and similarly energy intakes of men are higher than those of women. This is not entirely explained by their heavier weight, for per kilogram of body weight males still take more energy than females. So far as energy expenditure is concerned, a comparison of total energy expenditure of large numbers of

---

* Brown Adipose Tissue

males and females of any age over a sufficiently long time has, as far as I know, never been made. The literature on basal metabolism is large and often contradictory. On a surface area basis, women appear to have a lower basal metabolic rate than men, but when the results are recalculated to a body weight basis, the difference between the sexes is reduced (Durnin, 1976). Fat is the greatest variable in the human body, both in quality and quantity, but particularly in quantity, and those of us concerned with body composition know full well that the only way to get any sort of uniformity is to express the results for all the non-fat constituents of the body in terms of the fat-free body mass. As Durnin (1976), however, has pointed out, the fat-free mass is a mixture of metabolically active and inactive tissue, and adipose tissue is not inert. Liver and muscle, and above all brain, have a higher rate of oxygen consumption than adipose tissue, but the skeleton has a lower one. Even if the basal metabolic rates of males and females are more similar when expressed per kg of lean body mass than per kg or total body weight (Miller and Blyth, 1953; von Döbeln, 1956), lean body mass does not necessarily have the same proportions of more and less metabolically active tissue in the two sexes.

Table I shows the mean results of a study of Durnin et al. (1974) on the 7 day weighed food intakes of about 100 boys and 100 girls, all 14-15 years old in 1964, and of 200 of each sex in 1971. On both occasions, the boys took more energy than the girls, whichever basis of comparison is used. Durnin (1976) has suggested that males are more active than females, not only physically, but also metabolically, and this may be due to the influence of endocrine function, possibly by the thyroid, on metabolic activity.

Another point suggested by the results of Durnin et al. (1974) is that between 1964 and 1971, there was a reduction in the mean energy intake of both boys and girls. The authors attribute this to a lessening of physical activity between 1964 and 1971, which is probably part of a steady change over a longer period. More recently, Whitehead et al. (1982) have collected together the results of 16 studies on the energy intakes of boys and girls aged between one and 18 years, measured between 1930 and 1971 and have divided them into two groups, those made between 1930 and 1955 and those made between 1956 and 1971. In both earlier and later periods, the older boys

Table I : Mean body measurements and energy intakes
of 14 year old boys and girls (Durnin et al., 1974)

	Boys		Girls	
	1964	1971	1964	1971
Number	102	198	90	221
Height, m.	1.63	1.63	1.59	1.59
Weight, kg	51.1	50.8	51.8	50.7
Fat, g/100 g	16.3	18.4	27.6	27.0
Fat-free weight, kg	42.8	41.5	37.5	37.0
Surface area, sq.m.	1.51	1.50	1.52	1.50
Energy intake, kcal/day				
Total	2705	2610	2270	2020
Per kg body weight	52.9	51.4	43.8	39.8
Per kg fat-free weight	63.2	62.9	60.5	54.6
Per sq.m. surface area	1791	1740	1490	1347

took more energy than the girls, but it also appears that at all ages, girls in the more recent studies took less energy than they did in the earlier years. For boys, the difference becomes apparent at about 12 years. Body weights were similar in all the studies, and again a lessening of physical activity is put forward as the explanation of the decrease in energy intake over the years, though if this is the case the authors were puzzled as to why the younger boys had not altered.

One thing Whitehead et al. (1982) did not take into account was how the calorie intakes were arrived at in the various dietary surveys. I can only speak for those made in the United Kingdom where in most of the studies our own food tables were used, and in these tables the factors for calculating the energy value of foods from the protein, fat and carbohydrate have changed since the first edition was published in 1940 (McCance & Widdowson, 1940, 1942, 1946, 1960; Paul & Southgate, 1978). In all editions up to the fourth Rubner's factors were used. His factor of 17.15 kJ (4.1 kcal) per g carbohydrate was intended to be applied to a mixture of starch and sugar, but from the outset, we expressed all our carbohydrate values as monosaccharides. We therefore, in our ignorance in the 1930's, made an error in using the factors 17.15 (4.1); the correct energy value of 1 g monosaccharide, 15.69 kJ (3.75 kcal), was used in all editions after the first. All the calculations of the energy value of the diets of men (Widdowson, 1936), women (Widdowson & McCance, 1936) and children (Widdowson, 1947) were, however, based on the first edition of the tables and, in fact, many of them were made long before the tables were published. The absolute values given for energy intakes are, therefore, all too high.

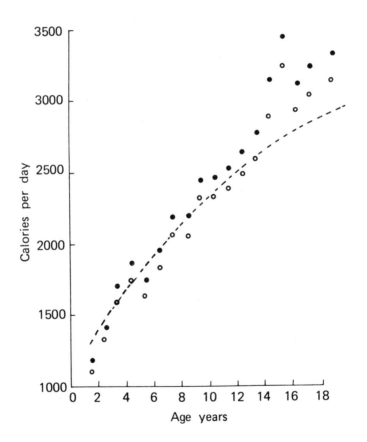

Fig. 1 : Energy intakes of boys - ● = Original values (Widdowson, 1947); o = recalculated using Atwater's factors (Widdowson, 1947); dotted line = fitted quadratic regression line for energy intakes reported in surveys after 1955 (Whitehead et al., 1982).

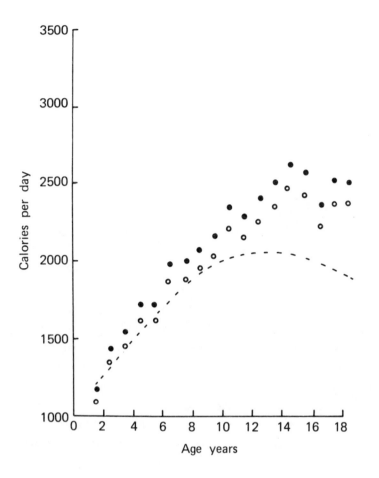

<u>Fig. 2</u> : Energy intakes of girls - ● = Original values (Widdowson, 1947);
o = recalculated using Atwater's factors (Widdowson, 1947); dotted line =
fitted quadratic regression line for energy intakes reported in surveys
after 1955 (Whitehead et al., 1982).

Figures 1 and 2 show the energy intakes of the boys and girls in my investigation (Widdowson, 1947) calculated from their mean intakes of protein,
fat and carbohydrate, first applying the factors I originally used as given
in the first edition of the food tables, and second according to the factors
in the fourth edition. At all ages, the later factors reduce the energy
intakes by about 10% and this is sufficient to bring the younger girls into
line with the Whitehead et al. (1982) regression line for energy intakes as

reported in dietary surveys after 1955. For boys from 14 years upwards and for girls from 10 years, there is still a large difference between the energy intakes 40 years ago and those within the period 1955 to 1971.

Why older children, particularly girls, should have been eating less over the years, but not younger ones, is a puzzling question, but the results of Durnin et al. (1974) would fit in with this.

Energy requirements in pregnancy

Another question that is exercising nutritionists at the present time concerns the energy of pregnant women. Do they or do they not eat more than they did before they were pregnant ? A workshop sponsored by Nestlé Nutrition was held at the Château de Rochegude, Vaucluse, France, in 1980 to discuss this problem (Dobbing, 1981) and an earlier one in Lutry, Switzerland, in 1979, sponsored by the Nestlé Foundation, touched on it (Aebi & Whitehead, 1980). Theoretically, women need more energy, at any rate during the last 2 or 3 months of pregnancy, but some investigators have found that they do not in fact increase their energy intakes at all. Naismith (1981) believes that the demands of the foetus for energy and nutrients are met by metabolic economies within the mother's body and he has provided evidence that this is indeed so in the case of protein. Animal studies suggest that protein is retained within the mother's body during the early part of pregnancy and this is withdrawn for the fetus during the latter part. During the last trimester of human pregnancy, the urinary excretion of 3-methylhistidine rises. Since this amino acid cannot be reutilized, a rise in excretion indicates that an increased catabolism of muscle protein is taking place. It is suggested that the other amino acids in the protein catabolized are being used for foetal growth. So far as total energy is concerned, the demands the foetus makes on the mother during the last weeks of gestation are somewhere between 100 and 300 kcal a day, about 30% of which is devoted to the synthesis of fat, 5% to the synthesis of protein and 65% to maintenance (Widdowson, 1981). It may be that any increase in a woman's energy intake during pregnancy is within experimental error of measurement. With the support of the Nestlé Foundation, further evidence on this is being sought (Durnin et al., 1982a; Durnin et al., 1982b). I must point out, however, that a proper comparison of a woman's pre-pregnant and pregnant food and energy intakes is rarely made. Either the "non-pregnant" intake is measured after the woman is known to be pregnant - in the new studies to which I have referred, the first measurement will be made between the 6th and 12th week, by which time the woman may have already increased her food intake to provide for the deposition of fat within her own body - or the pregnant intake is compared with that after pregnancy and lactation are over. Neither of these comparisons is strictly fair.

Energy cost of growth

My final point concerns the energy cost of growth. This is sometimes assumed to be the sum of the energy equivalent of the protein and fat stored in the body over a stated period of time and the energy cost of converting the absorbed nutrients into new body tissue. Taking the second part first, Millward et al. (1976) calculated that to convert fat into fat requires 0.01 kJ (0.002 kcal) per kJ deposited, carbohydrate into fat and protein into protein 0.15 kJ (0.036 kcal) per kJ, and protein into fat 0.31 kJ (0.071 kcal) per kJ. Pullar & Webster (1977) have simplified the whole thing by concluding that the energy costs of depositing 1 gram of protein or of fat are almost identical at 53 kJ (12.7 kcal) per gram. Schutz (1979) has also made some proposals, 46 kJ (11 kcal) per g fat tissue and 9.2 kJ (2.2 kcal) per g lean. Although these factors are expressed in different ways, they are in practice satisfactorily similar. Now, we come to the energy retained or stored. Schutz (1979) and others before him, including myself, have calculated this from the energy that would be produced if the protein and fat deposited in the body were incinerated in a bomb calorimeter, 23.8 kJ (5.7 kcal) per g protein and 38.9 kJ (9.3 kcal) per g fat. But they have not been derived from carbon, oxygen, hydrogen and nitrogen, but from protein, fat and carbohydrate in the food, so that to apply the bomb calorimeter factors to them seems quite inappropriate. The only additional energy required by the child for growth over and above the cost of converting absorbed nutrients to new body tissue seems to me to be the energy cost of processing the food in the digestive tract, and absorbing the products of digestion. I would welcome discussion on this point.

CONCLUSION

Although we know a great deal about how much food man requires, there still remain unsolved problems, some old, some new, most of them concerning differences between individuals, between the sexes, and between different ages. The old question posed by me 35 years ago "Why can one person live on half the calories of another, and yet remain a perfectly efficient physical machine ?" (Widdowson, 1947) has never been satisfactorily answered. Whether dietary-induced thermogenesis comes into it is controversial. Why males on the average need more energy than females, whatever basis of comparison is used, still remains open to investigation. Whether women require more energy after they become pregnant than they did before, and if not why not ?, is another topical subject.

The fall in energy intakes and, therefore, presumably requirements over the past 40 years is often attributed to decreasing physical activity but, if this is so, why do children before puberty not shown such a decrease ? One would suppose that school buses and television would have reduced their activity. Finally, the method of calculating the total energy cost of growth has not been completely settled. In a way, it is comforting to know that there are still problems to be solved 200 years after Lavoisier's experiments on the energy metabolism of himself and his guinea-pig.

REFERENCES

Aebi, H. & Whitehead, R. (Editors) (1980) Maternal Nutrition during Pregnancy and Lactation : A Nestlé Foundation workshop, Lausanne, April 26-27, 1979. Nestlé Foundation publication, series No 1. Hans Huber Publishers, Bern, Stuttgart, Vienna.

Bestley, J.W., Bramley, P.N., Dobson, P.M.S., Manhanty, A. & Tobin, G. (1982) Energy balance in "cafeteria"-fed young "Charles River" Sprague-Dawley rats. Proc. Physiol. Soc., 330:70.

Dallosso, H.M. (1982) An analysis of the causes of energy imbalance. Ph.D. Thesis, Cambridge University, Cambridge, England.

Darke, S., Disselduff, M. & Try, G.P. (1980) Frequency distributions of mean daily intakes of food energy and selected nutrients obtained during nutrition surveys of different groups of people in Great Britain between 1968 and 1971. Br. J. Nutr. 44:243-252.

Dobbing, J. (Editor) (1981) Maternal Nutrition in Pregnancy - Eating for Two ? A Nestlé Nutrition workshop, Vaucluse, France, June 1-4, 1980. Academic Press, London, New York.

Döbeln, W. von (1956) Human standard and maximal metabolic rate in relation to fat free body mass. Acta Physiol. Scand. 37:supplement 126.

Durnin, J.V.G.A. (1976) Sex differences in energy intake and expenditure. Proc. Nutr. Soc. 35:145-154.

Durnin, J.V.G.A. & Passmore, R. (1967) Energy Work and Leisure. Heinemann Educational Book Ltd., London.

Durnin, J.V.G.A., Lonergan, M.E., Good, J. & Ewan, A. (1974) A cross sectional nutritional and anthropometric study, with an interval of 7 years, on 611 young adolescent schoolchildren. Br. J. Nutr. 32:169-179.

Durnin, J.V.G.A., Hautvast, J., Valyasevi, A. & Whitehead, R.G. (1982a). Studies on the energy requirements of human pregnancy and lactation. Proc. Nutr. Soc., 41:145A.

Durnin, J.V.G.A., McKillop, F.M., Grant, S. & Fitzgerald, G. (1982b) Studies on the energy requirements of human pregnancy and lactation, . Proc. Nutr. Soc. 41:146A.

Edholm, O.G., Adam, J.M., Healy, M.J.R., Wolff, H.S., Goldsmith, R. & Best, T.W. (1970) Food intake and energy expenditure of army recruits. Br. J. Nutr. 24:1091-1107.

Edholm, O.G., Fletcher, J.G., Widdowson, E.M. & McCance, R.A. (1955) The energy expenditure and food intake of individual men. Br. J. Nutr. 9:286-300.

FAO/WHO/UNU (1983) Expert Consultation on Energy and Protein Requirements, in press.

Hervey, G.R. & Tobin, G. (1982) The part played by variation of energy expenditure in the regulation of energy balance. Proc. Nutr. Soc. 41:137-153.

James, W.T.P. & Trayhurn, P. (1981) Thermogenesis and obesity. Br. Med. Bull. 37:43-48.

Jung, R.T., Shetty, P.S. & James, W.P.T. (1980) Nutritional effects on thyroid and catecholamine metabolism. Clin. Sci. 58:183-191.

McCance, R.A. & Widdowson, E.M. (1940) The Chemical Composition of Foods. Medical Research Council Special Report Series No 235, HMSO, London.

McCance, R.A. & Widdowson, E.M. (1942) The Chemical Composition of Foods. Medical Research Council Special Report Series No 235, 3rd impression, revised, HMSO, London.

McCance, R.A. & Widdowson, E.M. (1946) The Chemical Composition of Foods. Medical Research Council Special Report Series No 235, 2nd edition, HMSO, London.

McCance, R.A. & Widdowson, E.M. (1960) The Composition of Foods. Medical Research Council Special Report Series No 297, 3rd revised edition of Special Report No 235, HMSO, London.

McCracken, K.J. & Barr, H.G. (1982) No impairment in efficiency of energy utilization in young growing rats kept at 24° and offered a varied diet. Proc. Nutr. Soc. 41:63A.

Miller, A.T. & Blyth, C.S. (1953) Lean body mass as a metabolic reference standard. J. Appl. Physiol. 5:311-316.

Miller, D.S. & Mumford, P. (1967) Gluttony. 1. An experimental study of over-eating on low or high protein diets. Am. J. Clin. Nutr. 20:1212-1222.

Miller, D.S. & Wise, A. (1975) Exercise and dietary-induced thermogenesis. Lancet i:1290.

Millward, D.J., Garlick, P.J. & Reeds, P.J. (1976) The energy cost of growth. Proc. Nutr. Soc. 35:339-349.

Morgan, J. & Mumford, P. (1977) Feeding practices and food intake of children under two years of age. Proc. Nutr. Soc. 36:47A.

Morgan, J., Mumford, P., Evans, E. & Wells, J. (1976) Food intake and anthropometric data on children under 4 years old living in the south of England. Proc. Nutr. Soc. 35:74A-75A.

Naismith, D.J. (1981) Diet during pregnancy - a rationale for prescription. In : Maternal Nutrition in Pregnancy - Eating for Two ? (Ed. Dobbing, J.) A Nestlé Nutrition workshop, Vaucluse, France, June 1-4, 1980. Academic Press, London, New York, pp. 21-40.

Neumann, R.O. (1902) Experimentelle Beiträge zur Lehre von dem täglichen Nahrungsbedarf des Menschen unter besonderer Berucksichtigung der notwendigen Eiweissmenge. Archiv. Hyg. Bakteriol. 45:1-87.

Passmore, R. (1982) Reflexions on energy balance. Proc. Nutr. Soc. 41: 161-165.

Paul, A.A. & Southgate, D.A.T. (1978) McCance & Widdowson's The Composition of Foods. Medical Research Council Special Report Series No 297, 4th edition, HMSO, London.

Pullar, J.D. & Webster, A.J.F. (1977) The energy cost of fat and protein deposition in the rat. Br. J. Nutr. 37:355-363.

Rothwell, N.J. & Stock, M.J. (1979) A role for brown adipose tissue in diet-induced thermogenesis. Nature (London) 281:31-35.

Rothwell, N.J. & Stock, M.J. (1982) Energy expenditure of "cafeteria"-fed rats determined from measurements of energy balance and indirect calorimetry. J. Physiol. 328:371-377.

Schutz, Y. (1979) Estimates of the energy cost of growth in young children. Int. J. Vitam. Nutr. Res. 49:113-124.

Whitehead, R.G., Paul, A.A. & Cole, T.J. (1982) Trends in food energy intakes throughout childhood from one to 18 years. Hum. Nutr. Appl. Nutr. 36:57-62.

Widdowson, E.M. (1936) A study of English diets by the individual method. Part I, Men. J. Hyg. 36:269-292.

Widdowson, E.M. (1947) A Study of Individual Children's Diets. Medical Research Council Special Report Series No 257, HMSO, London.

Widdowson, E.M. (1981) The demands of the fetal and maternal tissues for nutrients, and the bearing of these on the needs of the mother to "eat for two". In : Maternal Nutrition in Pregnancy - Eating for Two ? (Ed. Dobbing, J.) A Nestlé Nutrition workshop, Vaucluse, France, June 1-4, 1980. Academic Press, London, New York, pp. 1-19.

Widdowson, E.M. & McCance, R.A. (1936) A study of English diets by the individual method. Part II, Women. J. Hyg. 36:293-309.

THERMOGENIC RESPONSES INDUCED BY NUTRIENTS IN MAN :
THEIR IMPORTANCE IN ENERGY BALANCE REGULATION

Eric JEQUIER

Institute of Physiology, Division of Clinical Physiology,
University of Lausanne, Lausanne, Switzerland

SUMMARY

The regulation of body weight depends upon the control of
food intake and the regulation of energy expenditure. In man,
the control system for food intake may be overwhelmed by psychological or social influences and the thermogenic response to a
variable energy input may play an important role in the energy
regulatory system. Energy expenditure can be divided into
3 components : basal metabolic rate, thermogenesis and physical
activity. Of these 3 components, thermogenesis, (i.e. the energy
expended above the metabolic rate in the resting state) is the
most likely candidate to play a role in the regulation of energy
expenditure. The two main factors which contribute to thermogenesis, i.e. food intake and cold exposure, elicit diet-induced
thermogenesis (DIT) and non-shivering thermogenesis (NST),
respectively. It is of interest to study thermogenesis in individuals who present a tendency to gain weight, in order to
assess whether the thermogenic responses may be lower in these
subjects than in lean controls.

It has recently been shown that DIT consists of two separate components which can be described as "obligatory" and "regulatory" thermogenesis. The former is due to the energy costs of digesting, absorbing and converting the nutrients to their respective storage forms. The latter is an energy dissipative mechanism, mainly studied in animals. There is good experimental evidence showing that brown adipose tissue (BAT) is involved in the adaptive thermogenesis observed in rats fed a varied and palatable "cafeteria" diet. In addition, a thermogenic defect in BAT has been demonstrated in adult as well as young genetically obese animals, and this defect is present not only in adult, but also in young (12 day old) ob/ob mice, i.e. before the development of obesity. Thus, a defective thermogenesis seems to be a cause, rather than a consequence, of obesity in these animals.

In man, the role of thermogenesis in energy balance regulation is not yet understood. Some conflicting results may have arisen from inadequate techniques to measure energy expenditure. In our laboratory, we have developed three different techniques to measure energy expenditure in man, namely direct calorimetry, indirect calorimetry using an open-circuit ventilated hood system, and a respiratory chamber. Data from recent studies on DIT in man support the concept that a defect in thermogenesis may contribute to energy imbalance and weight gain in obese individuals.

We have been undertaking studies in obese subjects in an attempt to assess whether insulin resistance, which frequently occurs in obesity, affects glucose-induced thermogenesis (GIT). In 55 obese individuals studied over 3 hours following a 100 g oral glucose load, GIT was significantly reduced when compared to that of a control group. It is noteworthy that the magnitude of the reduction in GIT was related to the degree of insulin resistance; the more insulin-resistant patients had the lowest GIT. The measured value of GIT in these insulin-resistant patients corresponded well with the "obligatory" thermogenesis

of glucose. By contrast, control subjects exhibited a thermogenic response twice as great, suggesting both "obligatory" and
"regulatory" components in their GIT. These results suggest that
insulin may be required for the full DIT response, and that
insulin resistance contributes to blunt the "regulatory" thermogenesis. In this study we have shown that age also contributes
to a decrease in GIT. Thus both insulin resistance and age are
factors which decrease the thermogenic response to glucose.
These thermogenic defects may account, at least partially, for
the increasing occurrence of obesity with age, since they favour
energy retention and weight gain.

It is not yet established whether the thermogenic responses
due to lipid and protein ingestion are solely accounted for by
their respective "obligatory" thermogenesis. Studies on protein
turnover rates are needed in man to establish whether adaptive
modulations of synthesis and breakdown of protein contribute to
a variable thermogenesis.

Finally, it is noteworthy that in rats and mice, there are
many common features between DIT and NST including increases in
metabolic rate, in thermogenic response to noradrenaline and BAT
hypertrophy and hyperplasia. Moreover, these features are all
defective in genetically obese mice. In human obese female
individuals, we have also shown evidence for a deficiency in
both DIT and NST. Further studies are needed to establish
whether a defective thermogenesis may precede the development of
obesity in man, or whether it is a consequence of the increased
body fat mass. Our data showing the progressive decline in GIT
with increasing insulin resistance in the obese may favour the
latter alternative; however, more work is needed to study the
obese after weight loss with the possibility of detecting a
defective thermogenesis, independent of the weight excess, and
which may be genetically determined.

* * *

CONTROL OF FOOD INTAKE AND MODULATION OF ENERGY EXPENDITURE
IN THE REGULATION OF BODY WEIGHT

In the adult individual, stability of body weight results from a control of food intake and a modulation of energy expenditure. The two components of energy balance, i.e. input and output are closely interrelated. Workers involved in heavy physical activity obviously exhibit an elevated energy expenditure accompanied by a correspondingly increased food intake. Conversely, patients with anorexia nervosa, characterized by a very low food intake exhibit a low energy expenditure (Jéquier et al., 1978). Thus, both energy input and energy output influence each other (Garrow, 1978a).

It is generally admitted that the control of energy intake is the major factor in the regulation of energy balance. Since the control of food intake has been much studied, especially in the rat (Novin et al., 1976), it is not intended to discuss this topic here. It is, however, likely that in man, the regulatory system for food intake can be overwhelmed by psychological or social influences. Therefore, the degree of stability in body weight (or more precisely in body energy) will be largely dependent upon the individual's capacity to modulate energy expenditure so as to compensate for the variable energy intake.

Obesity has become a frequent condition in developed countries, indicating that the regulatory system of energy balance often works imperfectly in man. While it is often assumed that hyperphagia is the primary cause of obesity, most studies in man have failed to demonstrate an abnormal food intake in obese patients (Garrow, 1978b). Although assessment of food intake can be invalidated by various factors, such as changes in eating behaviour of the subjects while being studied and the difficulty in knowing the precise energy content of the ingested food, it is well-recognised that hyperphagia is not the unique cause of obesity in man and that other factors are involved.

This paper is divided into two parts, the first of which deals with the thermogenic responses to nutrient intake in man. It is of interest to compare the experimental data with the theoretical energy cost of absorbing and converting the nutrients to their respective storage forms. This comparison will show whether the measured thermogenic responses can be entirely predicted from the obligatory energy requirements for the biochemical processing of the nutrients, or whether other energy dissipative mechanisms may play a role. The second part of the paper describes recent evidence indicating that, in man, obesity may be due in part to defective thermogenic mechanisms.

THERMOGENIC RESPONSES TO NUTRIENT INTAKE

Responses to carbohydrate administration

After the ingestion of a meal, the energy expenditure increases. Many expressions have been used to describe this phenomenon, including specific dynamic action, postprandial thermogenesis, thermic effect of food and diet induced thermogenesis. The expression which is most used nowadays (Rothwell and Stock, 1981b) is "diet-induced thermogenesis" (DIT). Instead of reviewing the literature on DIT in man, this presentation will mainly deal with recent data obtained with our technique of computerized indirect calorimetry (Jéquier, 1981).

Diet-induced thermogenesis has long been considered as the obligatory energy costs of digesting, absorbing and processing or storing the nutrients. For carbohydrate, this "obligatory thermogenesis" depends upon the metabolic fate of the nutrient : when ingested glucose is absorbed and then directly oxidized, the increase in energy expenditure (i.e. the thermogenesis) only represents about 1%, whereas the cost of converting glucose to glycogen corresponds to about 5% of the glucose energy content (Flatt, 1978). Another pathway, lipogenesis from glucose, is energetically wasteful since the equivalent of 24% of the energy content of glucose converted into glycogen is expended (Flatt, 1978).

The experimental value for the thermogenesis induced by glucose ingestion in young adult individuals was found to be about 9% of the glucose energy ingested (Golay et al., 1982a), a value which could be accounted for by the three metabolic pathways, oxidation, glycogen synthesis and lipogenesis, provided that each one contributed similarly to glucose disposal. Since lipogenesis is quantitatively less important than the two other processes (Acheson et al., 1982), it can be inferred that the measured glucose induced thermogenesis is larger than the obligatory energy cost of glucose metabolism (Table I).

When glucose is administered by the intravenous route, together with insulin in order to maintain euglycemia (hyperinsulinaemic glucose clamp), about 85% of the infused glucose is taken up by muscles (DeFronzo et al., 1981), and less than 5% by the liver. Lipogenesis from glucose, being primarily a hepatic process in man, can only account for a negligible fraction of the infused glucose. By using different rates of glucose/insulin infusion, the glucose-induced thermogenesis was found to be lower than that obtained after oral administration of glucose (Thiébaud et al., 1982), but it was still larger than the obligatory energy cost of glucose metabolism, particularly when insulinaemia was elevated (Table I). These data support the concept that energy dissipative processes distinct from obligatory thermogenesis are induced in man by glucose administration.

Table I : Comparison between predicted and measured thermogenesis induced by nutrients in man[1]

Nutrient	Route of administration	Thermogenesis predicted	measured	References
Glucose	per os	4-5%	9%	Golay et al., 1982a
Glucose	intravenous[2]	4%	5.5 to 8%	Thiébaud et al., 1982
Carbohydrate (overfeeding)	per os	5-10%	27%	Schutz et al., 1982a
Fat	intravenous	2-3%	2-3%	Thiébaud et al., 1983
Fat + glucose	intravenous[2]	3%	5.5%	Thiébaud et al., 1983
Amino acids	per os[3]	25%	16.5%	Pittet et al., 1974

(1) The values reported in this table were obtained by our group; it is not intended to review values in the literature.
(2) During euglycemic hyperinsulinaemic glucose clamps.
(3) Measurement during 150 min. only.

Up to now, we have discussed acute responses to glucose administration over a period of a few hours. It is of major interest to know the magnitude of thermogenesis which can occur when energy intake is chronically increased above maintenance energy requirements. In a recent study of overfeeding with carbohydrates in our laboratory (the subjects received 1,500 kcal in excess of the preceding day's energy expenditure over a period of 7 days), Schutz et al., (1982a) observed an increase in energy expenditure corresponding to 27% of the excess energy intake (Table I). This value is much higher than the obligatory thermogenesis due to carbohydrate metabolism and indicates that a chronic excess of energy intake in man stimulates energy dissipative processes which tend to limit the gain of body energy. This "adaptive thermogenesis" may represent an important mechanism in body-weight regulation.

Thermogenic responses to fat and protein intake

Thermogenic responses to fat and to protein have been less studied in man than the effect of carbohydrate administration. In a recent study (Thiébaud et al., 1983), Intralipid (fat 200 g/l, glycerol 25 g/l, lecithin 12 g/l) infusion at a rate of either 0.23 mg/min. or 0.12 mg/min. was given to 7 young healthy volunteers. This lipid infusion induced a thermogenic response of 2 to 3% of the energy content of the lipid infused (Table I). This value is consistent with the theoretical energy cost for metabolising and storing lipid. Therefore, intravenous administration of lipids to man does not elicit energy dissipative processes distinct from obligatory thermogenesis. When, however, lipids were administered together with glucose and insulin (Thiébaud et al., 1983) the thermogenic response was found to be enhanced (Table I); it is likely that the thermogenesis induced by lipid then exceeds the value of the obligatory cost of lipid metabolism.

The thermogenic response to protein administration is classically described as greater than that of carbohydrate and lipid. Data on this topic in man are, however, rare and often difficult to interpret. The thermogenic response to protein (or amino acids) administration is long-lasting, and therefore is difficult to measure completely. The ingestion of 50 g of a balanced mixture of amino acids by young male volunteers produced a stimulation of thermogenesis of 16.5% of the energy content of the load when measured during 150 min. following the ingestion (Pittet et al., 1974). It is evident, however, that the thermogenic response was not completed after 150 min. and that the overall response is larger than that reported (Table I). In a recent study, Welle et al. (1981) compared the thermogenic response in man of equicaloric amounts of glucose, fat and protein; they reported a greater increment in energy expenditure after protein ingestion than after the intake of glucose or fat, but their values were curiously low.

The theoretical value for the thermogenic response to protein administration is large and amounts to about 25% of the energy content of the load (Flatt, 1978). This important increase in thermogenesis has been attributed to a stimulation of protein turnover rate and to the high energy cost of peptide bond synthesis. A comparable increase in energy expenditure is, however, elicited when ingested amino acids are oxidized, because of the high cost of ureogenesis and gluconeogenesis (Flatt, 1978). The reported values for thermogenesis induced by protein (or amino acid) ingestion are somewhat lower than the predicted ones, but it is likely that the total response has never been entirely measured because of the long duration of the process. These considerations lead to the concept that following protein (or amino acid) ingestion, the obligatory thermogenesis fully accounts for the increase in energy expenditure. When excessive amounts of proteins are ingested, energy dissipative mechanisms may, however, occur (Schutz, Y., personal communication).

In conclusion, the thermogenic response to food intake in man is dependent upon both the nature of the nutrients and the total amount of energy intake. The "obligatory thermogenesis" (i.e. the energy costs of digesting, absorbing and storing the nutrients) accounts for most of the thermogenic response to fat and protein administration. After carbohydrate ingestion, or after overfeeding, energy dissipative processes may contribute to the metabolic responses. The term "regulatory thermogenesis" is proposed for the increase in energy expenditure which is not accounted for by the "obligatory thermogenesis" after carbohydrate feeding, while the expression "adaptive thermogenesis" is used to describe the large increase in energy expenditure during overfeeding.

Mechanisms involved in energy dissipative processes after food intake

The mechanisms of the energy dissipative processes in man are still unknown. Increased lipid (Ball & Jungas, 1961) or protein (Yousef & Chaffer, 1970) turnover, substrate cycles in intermediary metabolism (Newsholme, 1980) or increased cation transport (Ismail-Beigi & Edelman, 1970) have been proposed to explain an increased utilization of ATP, but there is no convincing evidence supporting any of these mechanisms. Another possibility to account for an energy dissipative process is a decreased efficiency of ATP formation. The most compelling evidence for such a mechanism is the finding that the mitochondria of brown adipose tissue possess a unique pathway known as proton leakage or proton conductance pathway (Nicholls, 1979). This pathway allows the proton gradient generated by respiration to be dissipated, by permitting protons generated by respiration to move back across the inner mitochondrial membrane without the reaction being coupled to ATP production (Nicholls, 1979). Brown adipose tissue is found particularly in the interscapular and perirenal regions of the newborn and young animals and is considered to atrophy under normal conditions in adult animals, including man. This tissue has less triglyceride than white adipose tissue, is richly innervated with sympathetic nerve endings and has a very important vascular supply. The energy expenditure of this tissue can be enormously increased by sympathetic nerve stimulation or by injection of noradrenaline to the animal (Foster & Frydman, 1978). This tissue is known to be the major site of non-shivering thermogenesis in the rat (Foster & Frydman, 1978) and mouse (Thurlby & Trayhurn, 1980) during cold exposure.

It has been shown that modulations of the energy expended in brown adipose tissue also play an important role in preserving energy balance in rats (Rothwell & Stock, 1979). If rats are induced to eat a great amount of a highly palatable food ("cafeteria diet"), there is a large increase in thermogenesis which greatly reduces the weight gain predicted from the increased energy intake (Rothwell & Stock, 1979). It is surprising that a large part of this diet induced thermogenesis can be accounted for by an increased activity of brown adipose tissue (Rothwell & Stock, 1981a).

Thus in rodents, both conditions which stimulate thermogenesis, i.e. food intake and cold exposure, result in sympathetic activation of brown adipose tissue. The two processes, i.e. non-shivering thermogenesis (NST) and diet-induced thermogenesis (DIT), share many common features (Rothwell & Stock, 1980) including increase in food intake, stimulation of metabolic rate, enhanced thermogenic responses to noradrenaline and augmented brown adipose tissue thermogenesis. Brown adipose tissue has an important role in the new-born child since it is a major site of NST. Later in life, brown adipose tissue becomes atrophic and it was usually assumed that it disappeared. However, adult man still possesses some brown adipose tissue (Heaton, 1972) and there is indirect evidence that it may be functional (Rothwell & Stock, 1979); clearly this field needs to be investigated further by using techniques applicable to man.

Although there is no convincing evidence of brown adipose tissue activation in man after food ingestion, recent studies indicate that carbohydrate ingestion stimulates the activity of the sympathetic nervous system (Welle et al., 1981; Young et al., 1980). There is an acute increase in plasma noradrenaline levels following glucose administration (Welle et al., 1981; Young et al., 1980). Since plasma noradrenaline levels mainly represent an overflow of the neurotransmitter from sympathetic nerve endings, increased plasma levels of the transmitter is a good index of sympathetic nervous system activation. By contrast, ingestion of protein or fat did not significantly alter plasma noradrenaline levels (Welle et al., 1981).

The role of sympathetic stimulation in eliciting a part of the thermogenic response after glucose administration in man is further supported by the observation that infusion of propranolol (a drug which blocks β-receptors) significantly decreases the thermogenesis induced by glucose during hyperglycaemic hyperinsulinaemic glucose clamps (K.J. Acheson, personal communication). The residual increase in energy expenditure was fully accounted for by the predicted "obligatory thermogenesis". Thus, propranolol seems to inhibit most of the "regulatory thermogenesis", suggesting that the latter is mediated by activation of the sympathetic nervous system.

In rats, sucrose overfeeding is accompanied by an increase in noradrenaline turnover in the heart (Young & Landsberg, 1977), liver and pancreas

(Young & Landsberg, 1979), but the mechanisms of this stimulation are unknown. During euglycaemic hyperinsulinaemic glucose clamps in man, insulin infusion increases sympathetic nervous system activity (Rowe et al., 1981). This suggests the possibility of a central effect of insulin, or insulin-mediated glucose metabolism, on the activity of the sympathetic nervous system. This is particularly interesting since glucose (or carbohydrate) intake is accompanied by a rise in both insulin and noradrenaline plasma levels (Welle et al., 1981; Young et al., 1980). Thus, plasma insulin levels, considered as an index of insulin-mediated glucose metabolism within a small region of the central nervous system (perhaps in the hypothalamus) might be a signal linking carbohydrate intake to sympathetic activity. If the latter controls thermogenesis through brown adipose tissue or other effectors (Fig. 1), a link would have been established between energy intake and output. This might contribute to explain the "regulatory thermogenesis", i.e. the energy dissipative process induced by carbohydrate intake which is not accounted for by the energy costs of glucose metabolism.

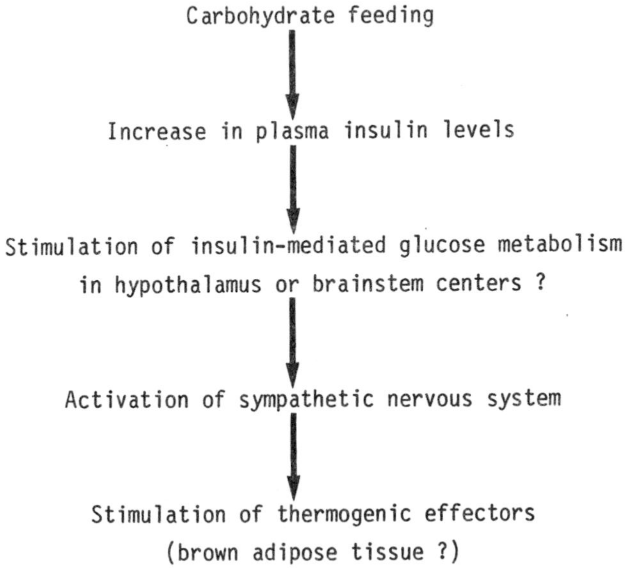

Carbohydrate feeding

↓

Increase in plasma insulin levels

↓

Stimulation of insulin-mediated glucose metabolism
in hypothalamus or brainstem centers ?

↓

Activation of sympathetic nervous system

↓

Stimulation of thermogenic effectors
(brown adipose tissue ?)

Fig. 1 : Model of the mechanism of "regulatory thermogenesis" after carbohydrate feeding.

Further studies are needed to explore the relationships between energy intake and output. In this context, it is of interest that the ventromedial hypothalamus (VMH) may affect both feeding and thermogenesis (Perkins et al., 1981). Electrical stimulation of the VMH area results in sympathetic activation of thermogenesis in brown adipose tissue (Perkins et al., 1981) while electrolytic or chemical lesions in the VMH induce hyperphagia and excess weight gain which indicates that this area may be involved in the control of food intake.

DO THERMOGENIC DEFECTS CONTRIBUTE TO THE ENERGY IMBALANCE
WHICH LEADS TO OBESITY IN MAN ?

The renewed interest in considering the putative role of thermogenesis in the regulation of energy balance stems from recent experimental data in animals. Two models are of particular interest, viz. :

1) hyperphagia which is not accompanied by obesity, and

2) obesity occurring without hyperphagia.

The first model is represented by weanling rats on a "cafeteria diet" which do not gain any excess weight in spite of a 50-60% increase in food intake (Rothwell & Stock, 1981b), and the second by hypothalamic and genetic obesity. In weanling rats, lesion of the ventro-medial hypothalamus induces an increase in body fat without any increase in food intake (Goldman et al., 1977). More compelling evidence that obesity can occur without hyperphagia results from studies in genetic obesity. In pair feeding studies, (ob/ob) mice (Thurlby & Trayhurn, 1979) become obese when receiving the same amount of food as their lean littermates. The propensity towards obesity of the mutants was found to be mainly due to a decreased thermogenesis resulting from a mitochondrial defect in brown adipose tissue (Himms-Hagen & Desautels, 1978; Thurlby & Trayhurn, 1980); the diabetic (db/db) mouse (Case & Powley, 1977) and the fatty rat (Clearly et al., 1980) also become obese when pair-fed with their lean littermates. It is interesting that these (ob/ob) and (db/db) mice also exhibit a defective non-shivering thermo-genesis (NST) when exposed to the cold. In the (ob/ob) mouse, the defect in

NST can already be shown in 12-day-old animals before the development of obesity (Trayhurn et al., 1977); this suggests that the thermogenic defect in these mutants is genetically determined and may be a cause rather than a consequence of obesity.

Whereas the evidence is very strong that a reduced thermogenesis (both NST and DIT) contributes to the development of obesity in genetically obese rodents (Himms-Hagen & Desautels, 1978; Thurlby & Trayhurn, 1979, 1980; Trayhurn et al., 1977) and in rats with lesion of the ventromedial hypothalamus (Seydoux et al., 1981), the putative role of a thermogenic failure in human obesity is more difficult to assess.

Obesity arises from a situation where energy intake chronically exceeds energy output, but it is not clearly established in man whether this imbalance results from an excessive input or a defective output. It is likely that the two extreme conditions, i.e. hyperphagia on one hand and subnormal thermogenesis on the other may only account for a few cases of obesity. The present understanding is that obesity results from an altered control system affecting in various degrees both energy intake and energy output; in some individuals hyperphagia may be the obvious cause of obesity, whereas in other obese subjects, a defective thermogenic capacity unable to adapt energy output to a variable intake, may have an important role favouring energy gain. According to this view, the aetiology of human obesity may be considered as a spectrum from obvious hyperphagia to subnormal thermogenesis. One major difficulty in studying human obesity is due to the fact that there is presently no adequate procedure to assess the relative importance of excess intake versus low output in a given patient. It is likely that most studies on human obesity are performed on heterogeneous populations, which may explain some of the discrepancies in the data obtained by different groups.

To study the role of energy output in energy balance, it is necessary to consider the 3 components of energy expenditure i.e. : basal metabolic rate, thermogenesis and physical activity.

Basal metabolic rate is mainly dependent upon the size of the fat free mass (i.e. body weight minus fat mass) (Halliday et al., 1979; Ravussin et al., 1982). Since obese individuals often have an increased fat free mass in addition to their large fat mass, their basal metabolic rate is usually more elevated than that of lean control subjects (James et al., 1978; Ravussin et al., 1982). Thermogenesis can be defined as the difference between resting energy expenditure and basal metabolic rate. It results from stimuli such as food intake, cold exposure, psychological influences, thermogenic agents, activity of the sympathetic nervous system, and hormonal influences. Lastly, physical activity is obviously a component which affects energy expenditure. In the developed countries, however, most adult individuals exhibit a range of "normal activity" which only represents 15 to 20% of the total energy expenditure (Garrow, 1978a). In addition, in sedentary individuals, physical activity of obese and lean subjects was found to be very similar (McCarthy, 1966; Schutz et al., 1982b; Stefanik et al., 1959).

In human obesity, thermogenic defects have been described (James & Trayhurn, 1976) and may play a role in the energy imbalance leading to excessive weight gain. Both dietary-induced thermogenesis (Jéquier et al., 1978) and non-shivering thermogenesis (Jéquier et al., 1974) are defective in cases of familial obesity.

In a recent study (Golay et al., 1982b), we have reported a decreased thermogenesis induced by glucose ingestion in obese subjects; this thermo-genic response was particularly reduced when insulin resistance was present. Furthermore, the lowest glucose induced thermogenesis was found in obese diabetics with a reduced insulin response to the glucose load. These results suggest that insulin is required for glucose-induced thermogenesis. As mentioned previously, insulin may also be a signal for the hypothalamus and may contribute to the activation of the sympathetic nervous system. The low thermogenic response in obese subjects with insulin resistance (or with a reduced insulin response) could be related to a lack of activation of the sympathetic system.

Changes in plasma noradrenaline levels after a meal have, however, been observed in obese subjects (Shetty et al., 1981); further work is needed to establish whether a blunted response of the sympathetic nervous system is responsible for the low thermogenesis observed in the obese. In addition, infusion of noradrenaline has been reported to elicit a lower thermogenic response in obese women than in lean controls (Jung et al., 1979); the response to noradrenaline was also decreased in "post-obese women" (i.e. obese after significant weight loss).

In elderly non-obese subjects, the thermogenesis induced by glucose ingestion was found to be lower than in young controls (Golay et al., 1982b). Thus, both age and obesity are factors which contribute to lowering the thermogenic response to glucose.

In conclusion, of the 3 components of energy expenditure, basal metabolic rate is usually augmented in obesity, thermogenesis is reduced, and physical activity (although variable) is often similar in the obese when compared to lean sedentary individuals. The total energy expended over a 24 h period was found to be somewhat larger in a group of obese individuals when compared to lean controls (Ravussin et al., 1982). Therefore, the decreased thermogenesis in the obese does not fully compensate for their elevated basal metabolic rate; the conclusion has been drawn from similar results obtained by Garrow (1981) that a failure in the thermogenic response to various stimuli cannot have any important role in causing obesity in human subjects. This concept, however, does not take into account the fact that a few obese individuals have a lower total energy expenditure than lean controls (Ravussin et al., 1982), which illustrates the heterogeneity of the energy metabolism in obese individuals. In addition, after weight loss, a few previously obese subjects had a reduced overall energy expenditure (E. Ravussin, personal communication).

Thus, the following hypothesis has been proposed (James & Trayhurn, 1976) : In individuals who have a propensity to obesity, a decreased energy output resulting from a failure in thermogenesis may favour a positive energy balance and subsequent weight gain. With the increase in fat mass, the lean body mass also becomes larger (although the increase in the latter is much less than that of the former), which results in an elevated basal metabolic rate. This eventually leads to a compensation of the low thermogenesis, and total body expenditure normalises or may even become somewhat larger than that of lean individuals. Since the thermogenic capacity is still defective, any excess in food intake will be preferentially stored, and body weight further increases.

More work is needed to assess the role of thermogenesis in body weight regulation in man : however, present evidence suggests that the control of thermogenesis may be as important as the control of food intake. The recommendation to eat thermogenic nutrients and the development of thermogenic agents may prove to be useful in the treatment of human obesity.

REFERENCES

Acheson, K.J., Flatt, J.P. & Jéquier, E. (1982) Glycogen synthesis versus lipogenesis after a 500 gram carbohydrate meal in man. Metabolism 31: 1234-1240.

Ball, E.G. & Jungas, R.L. (1961) On the action of hormones which accelerate the rate of oxygen consumption and fatty acid release in rat adipose tissue in vitro. Proc. Natl. Acad. Sci. USA 47:932-941.

Case, J.E. & Powley, T.L. (1977) Development of obesity in diabetic mice pair-fed with lean siblings. J. Comp. Physiol. Psychol. 91:347-358.

Clearly, M.P., Vasselli, J.R. & Greenwood, M.R.C. (1980) Development of obesity in the Zucker obese (fafa) rat in the absence of hyperphagia. Am. J. Physiol. 238:E284-E292.

DeFronzo, R.A., Jacot, E., Jéquier, E., Maeder, E., Wahren, J. & Felber, J.P. (1981) The effect of insulin on the disposal of intravenous glucose : results from indirect calorimetry and hepatic and femoral venous catheterization. Diabetes 30:1000-1007.

Flatt, J.P. (1978) The biochemistry of energy expenditure. In : Recent Advances in Obesity Research : II (Ed. G. Bray). Newman Publishing London, pp. 211-218.

Foster, D.O. & Frydman, M.L. (1978) Nonshivering thermogenesis in the rat. II. Measurement of blood flow with microspheres point to brown adipose tissue as the dominant site of the calorigenesis induced by noradrenaline. Can. J. Physiol. Pharmacol. 56:110-122.

Garrow, J.S. (1978a) The regulation of energy expenditure in man. In : Recent Advances in Obesity Research : II (Ed. G. Bray). Newman Publishing London, pp. 200-210.

Garrow, J.S. (1978b) Energy balance and obesity in man. Elsevier/North-Holland Biomedical Press, Amsterdam, New York, Oxford, p. 48.

Garrow, J.S. (1981) Thermogenesis and obesity in man. In : Recent Advances in Obesity Research : III (Ed. P. Björntorp, M. Cairella & A.N. Howard). J. Libbey, London, pp. 208-215.

Golay, A., Schutz, Y., Thiébaud, D., Curchod, B., Felber, J.P. & Jéquier, E. (1982a) Influence of insulin resistance on glucose induced thermogenesis in man. Experientia 38:714.

Golay, A., Schutz, Y., Meyer, H.U., Thiébaud, D., Curchod, B., Maeder, E., Felber, J.P. & Jéquier, E. (1982b) Glucose induced thermogenesis in non-diabetic and diabetic obese subjects. Diabetes 31:1023-1028.

Goldman, J.K., Bernadis, L.L. & Frohman, L.A. (1977) Food intake in hypo-thalamic obesity. Am. J. Physiol. 227:88-91.

Halliday, D., Hesp, R., Stalley, S.F., Warwick, P., Altman, D.G. & Garrow, J.S. (1979) Resting metabolic rate, weight, surface area and body composi-tion in obese women. Int. J. Obesity 3:1-6.

Heaton, J.M. (1972) The distribution of brown adipose tissue in the human. J. Anat. 112:35-39.

Himms-Hagen, J. & Desautels, M. (1978) A mitochondrial defect in brown adipose tissue of obese (ob/ob) mouse : reduced binding of purine nucleo-tides and a failure to respond to cold by an increase in binding. Biochem. Biophys. Res. Comm. 83:628-634.

Ismail-Beigi, F. & Edelman, I.S. (1970). Mechanisms of thyroid calori-genesis : role of active sodium transport. Proc. Natl. Acad. Sci. USA 67:1071-1078.

James, W.P.T. & Trayhurn, P. (1976) An integrated view of the metabolic and genetic basis for obesity. Lancet 2:770-772.

James, W.P.T., Bailes, J., Davies, H.L. & Dauncey, M.J. (1978) Elevated metabolic rates in obesity. Lancet 1:1122-1125.

Jéquier, E. (1981) Long term measurement of energy expenditure in man : direct or indirect calorimetry ? In : Recent Advances in Obesity Research : III (Ed. P. Björntorp, M. Cairella & A.N. Howard). J. Libbey, London, pp. 130-135.

Jéquier, E., Gygax, P.H., Pittet, P. & Vannotti, A. (1974) Increased thermal body insulation : relationship to the development of obesity. J. Appl. Physiol. 36:674-678.

Jéquier, E., Pittet, Ph. & Gygax, P.H. (1978) Thermic effect of glucose and thermal body insulation in lean and obese subjects : a calorimetric approach. Proc. Nutr. Soc. 37:45-53.

Jung, R.T., Shetty, P.S. & James, W.P.T. (1979) Reduced thermogenesis in obesity. Nature (London) 279:322-323.

McCarthy, M.G. (1966) Dietary and activity patterns of obese women in Trinidad. J. Am. Diet. Ass. 48:33-37.

Newsholme, E.A. (1980) A possible basis for the control of body weight. N. Eng. J. Med., 302:400-405.

Nicholls, D.G. (1979) Brown adipose tissue mitochondria. Biochim. Biophys. Acta 549:1-29.

Novin, D., Wyrwicka, W. & Bray, G. (Ed.) (1976) Hunger : Basic mechanisms and clinical implications. Raven Press, New York, 494 pp.

Perkins, N.M., Rothwell, N.J., Stock, M.J. & Stone, T.W. (1981) Activation of brown adipose tissue thermogenesis by the ventromedial hypothalamus. Nature (London) 209:401-402.

Pittet, Ph., Gygax, P.H. & Jéquier, E. (1974) Thermic effect of glucose and amino acids in man studied by direct and indirect calorimetry. Br. J. Nutr. 31:343-349.

Ravussin, E., Burnand, B., Schutz, Y. & Jéquier, E. (1982) Twenty-four-hour energy expenditure and resting metabolic rate in obese, moderately obese, and control subjects. Am. J. Clin. Nutr. 35:566-573.

Rothwell, N.J. & Stock, M.J. (1979) A role for brown adipose tissue in diet-induced thermogenesis. Nature (London) 281:31-55.

Rothwell, N.J. & Stock, M.J. (1980) Similarities between cold and diet-induced thermogenesis in the rat. Can. J. Physiol. Pharmacol. 58:842-848.

Rothwell, N.J. & Stock, M.J. (1981a) Influence of noradrenaline on blood flow to brown adipose tissue in rats exhibiting diet-induced thermogenesis. Pflügers Arch. 389:237-242.

Rothwell, N.J. & Stock, M.J. (1981b) Regulation of energy balance. Ann. Rev. Nutr. 1:235-256.

Rowe, J.W., Young, J.B., Minaker, K.L., Stevens, A.L., Pallotta, J. & Landsberg, L. (1981) Effect of insulin and glucose infusions on sympathetic nervous system activity in normal man. Diabetes 30:219-225.

Schutz, Y., Acheson, K., Bessard, T. & Jéquier, E. (1982a) Energy balance during one week carbohydrate overfeeding in man. Int. J. Vitam. Nutr. Res. 52:208.

Schutz, Y., Ravussin, E., Diethelm, R. & Jéquier, E. (1982b) Spontaneous physical activity measured by radar in obese and control subjects in a respiration chamber. Int. J. Obesity 6:23-28.

Seydoux, J., Rohner-Jeanrenaud, F., Assimacopoulos-Jeannet, F., Jeanrenaud, B. & Girardier, L. (1981) Functional disconnection of brown adipose tissue in hypothalamic obesity in rats. Pflügers Arch. 390:1-4.

Shetty, P.S., Jung, R.T., James, W.P.T., Barrand, M.A. & Callingham, B.A. (1981) Postprandial thermogenesis in obesity. Clin. Sci. 60:519-525.

Stefanik, P.A., Heald, F.P. & Mayer, J. (1959) Calorie intake in relation to energy output of obese and non-obese adolescent boys. Am. J. Clin. Nutr. 7:55-62.

Thiébaud, D., Schutz, Y., Acheson, K., Jacot, E., DeFronzo, R., Felber, J.P. & Jéquier, E. (1982) Thermogenesis induced by intravenous glucose/insulin infusion in healthy young men. Int. J. Vitam. Nutr. Res. 52:209.

Thiébaud, D., Acheson, K., Schutz, Y., Felber, J.P., Golay, A., DeFronzo, R. & Jéquier, E. (1983) Stimulation of thermogenesis in man following combined glucose-long-chain triglyceride infusion. Am. J. Clin. Nutr. 37:603-611.

Thurlby, P.L. & Trayhurn, P. (1979) The rôle of thermoregulatory thermo-genesis in the development of obesity in genetically obese (ob/ob) mice pair fed with lean siblings. Br. J. Nutr. 42:377-385.

Thurlby, P.L. & Trayhurn, P. (1980) Regional blood flow in genetically obese (ob/ob) mice. Pflügers Arch. 385:193-201.

Trayhurn, P., Thurlby, P.L. & James, W.P.T. (1977) Thermogenic defect in pre-obese ob/ob mice. Nature (London) 266:60-62.

Welle, S., Lilavivat, U. & Campbell, R.G. (1981) Thermic effect of feeding in man : increased norepinephrine levels following glucose but not protein or fat consumption. Metabolism 30:953-958.

Young, J.B. & Landsberg, L. (1977) Stimulation of the sympathetic nervous system during sucrose feeding. Nature (London) 269:615-617.

Young, J.B. & Landsberg, L. (1979) Effect of diet and cold exposure on norepinephrine turnover in pancreas and liver. Am. J. Physiol. 236:E524-E533.

Young, J.B., Rowe, J.W., Pallotta, J.A., Sparrow, D. & Landsberg, L. (1980) Enhanced plasma norepinephrine response to upright posture and oral glucose administration in elderly human subjects. Metabolism 29:532-539.

Yousef, M.K. & Chaffer, R.R.J. (1970) Studies on protein-turnover in cold-acclimated rats. Proc. Soc. Exp. Biol. Med. 133:801-804.

ENERGY EXPENDITURE AND WHOLE BODY PROTEIN SYNTHESIS
IN VERY LOW BIRTH WEIGHT (VLBW) INFANTS

Y. SCHUTZ*, C. CATZEFLIS*, F. GUDINCHET[+], J. MICHELI[+],
C. WELSCH, M.J. ARNAUD and E. JEQUIER*

Institute of Physiology* and Service of Paediatrics[+], Faculty of Medicine,
University of Lausanne, Switzerland, and

Nestlé Products Technical Assistance Co. Ltd., Research Department,
CH-1814 La Tour-de-Peilz, Switzerland

SUMMARY

To examine the rates of whole body protein synthesis and energy expenditure during the rapid growing period, premature infants of very low birth weight (VLBW) (< 1500 g), appropriate for gestational age were kept under standard thermoneutrality conditions and received a formula diet providing 110 kcal/kg.d metabolisable energy (ME) and 3.3 g protein/kg.d. Their energy expenditure was measured by open circuit indirect calorimetry. Nitrogen turnover and whole body protein synthesis and catabolism were determined using repeated oral administration of ^{15}N-glycine for 60-72 h followed by the analysis of ^{15}N-enrichment in urinary urea.

These VLBW infants grew at an average rate of 15 g/kg.d. About half of the ME intake (i.e. 50 kcal/kg.d) was invested in weight gain while the remainder (i.e. 60 kcal/kg.d) was oxidised. The energy equivalent of the weight gain (i.e. the

amount of energy stored per g weight gain) and the N balance
indicated that lean tissue made up approximately 2/3 of the
weight gained and fat tissue the remaining 1/3.

The plateau value for ^{15}N enrichment reached on the third
day of administration allowed us to calculate a rate of protein
synthesis of 14 g/kg.d and protein breakdown of 12 g/kg.d in
five VLBW fed a formula diet.

The elevated energy expenditure of the very low birth weight
infant seems to be related to its rapid rate of weight gain
which is accompanied by a high rate of body protein synthesis.
More than 20% of the total energy expenditure of the VLBW
infants was accounted for by whole body protein synthesis.

* * *

Over the years sophistication in calorimetric techniques has largely
contributed to the better understanding of the energy metabolism of very low
birth weight (VLBW) infants, i.e. infants with a birth weight below 1500 g.
These methods now make it possible to evaluate the metabolic impact of
various nutritional regimens under controlled clinical conditions. An examination of the rate of whole body protein synthesis and breakdown in VLBW
infants is of interest for two reasons : firstly, because protein turnover
may account for a large fraction of the resting energy expenditure;
secondly, because current knowledge of protein turnover is mainly restricted
to that of infants recovering from malnutrition (Waterlow et al., 1978).

In newborn children characterized by a rapid growth, protein accumulation
is a dynamic process which results from a simultaneous synthesis and
catabolism of proteins. While total protein gain can be calculated from
nitrogen balance studies, isotopic tracer techniques are necessary to
evaluate protein turnover, synthesis and catabolism in humans. Non-
radioactive, stable isotopes of low natural abundance - 1.11% for ^{13}C and
0.36% for ^{15}N, can be used to this end. Administration of molecules
enriched with stable isotopes will lead to a significant enrichment which

can be determined in expired gas, urine or blood. The metabolism of a molecule is generally unaffected by the enrichment with a stable isotope although significant isotopic effects have been reported to occur with ^{18}O and particularly deuterium (Thomson, 1963). In this study, repeated oral administration of ^{15}N-glycine was used to assess protein turnover rate (Picou & Taylor-Roberts, 1969).

Energy cost of weight gain

Healthy VLBW infants present a unique opportunity to study the energy cost of growth, since high rates of weight gain are achieved. The extrauterine gain between the 34th and 36th week of gestational age is approximately 15 g/kg.d (Largo et al., 1980), which is roughly 3 times the value measured for term-infants, and 30 times the value for healthy adolescents, using Swiss reference standards (Prader et al., 1980). Rapid weight gain is also found in children recovering from protein-energy deprivation (Spady et al., 1976; Golden et al., 1977). At the neonate Division of the Pediatric Clinic in Lausanne, VLBW infants are fed approximately 125 kcal/kg.d of gross energy (as determined by bomb calorimetry). When the energy losses in faeces and urine are taken into account, the metabolizable ("available") energy (ME) given to VLBW infants fed on a formula diet is approximately, 110 kcal/kg.d, i.e. less than 90% of the gross energy intake (unpublished results).

The resting energy expenditure of the VLBW infants was measured by continuous open circuit, indirect calorimetry under standard thermoneutrality conditions (Gudinchet et al., 1982). The measurements were made using an air tight perspex box, placed inside an incubator, and containing the infant. Air within the box was continuously renewed and the flow rate measured by a pneumotachograph connected to a differential manometer. The outflowing air was continuously sampled and its O_2 and CO_2 concentrations were measured using a mass spectrometer. The infant's oxygen consumption ($\dot{V}O_2$) and CO_2 production ($\dot{V}CO_2$) were determined from the flow rate and differences in O_2 and CO_2 concentrations between air entering and leaving the calorimeter. Energy expenditure was calculated from $\dot{V}O_2$, $\dot{V}CO_2$ and respiratory quotient (RQ).

The mean, resting postprandial energy expenditure of the VLBW infants was found to be 60 kcal/kg, between the 2nd and 7th weeks of life, whereas the value for term-babies is 48 kcal/kg.d (Widdowson, 1981). Energy retention (i.e. the difference between ME intake and energy expenditure) averaged therefore to 50 kcal/kg.d, indicating that about half (45%) of the ME intake was channelled to growth (Fig. 1). In situations where infants are growing very rapidly, i.e. in VLBW infants or those recovering from protein-energy deprivation, it is possible to divide their overall energy expenditure into growth and non-growth (or maintenance) functions (Fig. 1). In this study, the non-growth energy expenditure was not determined directly, but was derived from the "zero" weight gain value of the regression line of energy expenditure (kcal/kg.d) vs. weight gain (g/kg.d). With 48 neonates, such relationships were highly significant (r = 0.58, p < 0.001).

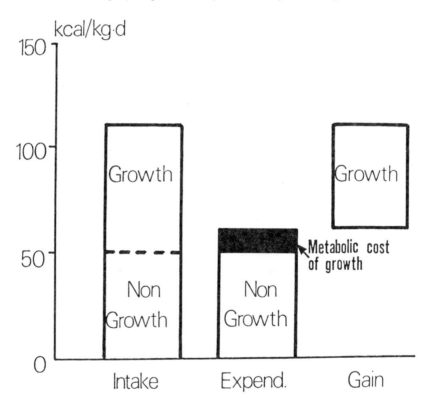

Fig. 1 : Partition of metabolisable energy intake (i.e. gross energy intake minus faecal and urine energy) and energy expenditure into growth and non growth functions in a very low birth weight gaining 15 g/kg.d.

The non-growth energy expenditure was found to be 51 kcal/kg.d (Gudinchet et al., 1982). To store body energy, an extra-amount of calories is expended above the non-growth energy expenditure. This is the metabolic cost of growth (Fig. 1), which represents an extra-thermogenesis for synthetizing new body tissue. There are several ways of assessing this compartment (Schutz, 1979). In this study, the metabolic cost of growth was calculated from the slope of the regression line of energy expenditure vs. weight gain, i.e. 0.54 kcal. per g weight gain (Gudinchet et al., 1982). This value is consistent with that found by other investigators studying VLBW infants with similar characteristics (Chessex et al., 1981).

Estimates of the composition of tissue gain

From the energy equivalent of the weight gain an index of the composition of the gain can be obtained (Schutz, 1979), using the following equation :

$$\frac{\text{ME intake (kcal/kg.d)} - \text{E Expended (kcal/kg.d)}}{\text{weight gain (g/kg.d)}}$$

$$\text{i.e.} \frac{110-60}{15} = 3.3 \text{ kcal/g weight gain.}$$

If one assumes that the lean tissue gain of VLBW infants has a protein content of 20% (i.e. an energy equivalent of 1.1 kcal/g) and that fat tissue gain has 85% lipid (i.e. an energy equivalent of 7.9 kcal/g), our results would indicate that the weight gain is made up of 68% lean and 32% fat tissues. An independent estimate of the composition of the tissue gain can be obtained from the N balance studies performed over 72 consecutive hours. The mean protein gain was found to be approximately 2 g prot/kg.d. Therefore, the lean tissue gain would be estimated to be 10 g/kg.d or 67% of the weight gain, a value in good agreement with the previous estimate.

Rates of whole body protein synthesis and breakdown

In a small number of VLBW infants, we examined the rates of whole body protein synthesis and breakdown during their rapid growth period (2nd to 7th postnatal week). The infants received a milk formula providing the same amount of energy given previously and 3.3 g protein per kg body weight per day.

The model used for in vivo analysis of total nitrogen turnover assumes a single pool of metabolic nitrogen, constant in size during the study (Waterlow et al., 1978). This pool receives amino acids from the food ($\dot{I}$) and from protein catabolism ($\dot{C}$). Amino acids are withdrawn from this pool for protein synthesis ($\dot{S}$) or, excreted in the urine as end-products of amino acid metabolism ($\dot{E}$). If this pool is assumed to be constant in size, the input of amino acid nitrogen is equal to the output and corresponds to $\dot{Q}$, which is the rate of total nitrogen flux :

$$\dot{I} + \dot{C} = \dot{S} + \dot{E} = \dot{Q} \qquad (1).$$

When $\dot{Q}$ is known, synthesis ($\dot{S}$) and catabolism ($\dot{C}$) of protein nitrogen can be calculated since the intake of nitrogen ($\dot{I}$) and its excretion ($\dot{E}$) can be analytically determined.

Infant formula was given every three hours, by intra-gastric route. ^{15}N-glycine was added for 60-72 hours at a concentration of 0.18 mg/ml, corresponding to a rate of administration of ^{15}N ($\dot{d}$) of 5 mg ^{15}N/kg.day. When the enrichment of the metabolic nitrogen pool had reached steady state, any products leaving this pool (protein synthesis and urea and ammonia excretion) would theoretically agree with the following relationship :

$$\frac{\dot{d}}{\dot{Q}} = \frac{\dot{e}_{urea}}{\dot{E}_{urea}} = \frac{\dot{e}_{NH_3}}{\dot{E}_{NH_3}} \qquad (2)$$

where the ratio of the rate of ^{15}N administered to the rate of total nitrogen flux is the same as the ratio of the rate of ^{15}N in urea or ammonia to the rate of excretion of total urea or ammonia in urine. These last two ratios are in fact the ^{15}N enrichment of excreted urea and ammonia. The ^{15}N enrichment was determined automatically by emission spectrometry (Isonitromat 5201, VEB Statron, DDR) from ammonium chloride

samples. These samples were prepared by trapping ammonia after alkaliniza-
tion of urine. For urea nitrogen analysis, the urine samples were further
neutralized, urease added and, following realkalinization, the ammonia
produced was trapped.

The nitrogen flux ($\dot{Q}$) was obtained from equation 2 and synthesis ($\dot{S}$) and
catabolism ($\dot{C}$) were calculated from equation 1. Urine samples were collected
every 3 hours for 72 hours. From the value of the plateau of ^{15}N enrichment in ammonia and urea, the N flux was calculated. The use of [15]N-
glycine is well-documented and similar results of protein synthesis have
been obtained in man, with ^{15}N-glycine and a uniformly ^{15}N enriched egg
protein (Picou & Taylor-Roberts, 1969).

In order to approximate the isotopic plateau a single exponential of the
form $Y = Y_0 (1-e^{-kt})$ and the visual inspection approach (Waterlow et
al., 1978) were also used. According to the last authors, the most
satisfactory approach seems to be visual inspection.

The kinetics of ^{15}N enrichment in urinary urea and ammonia, expressed
as At% [15]N excess, and obtained in one preterm infant is presented in
Fig. 2. This figure also shows urea enrichment reported previously in adult
subjects (Steffee et al., 1976) and preterm infants (Jackson et al., 1981).
Except for the low enrichment observed by Jackson et al. (1981), these
curves exhibit a plateau, or more precisely a pseudo-plateau because the
enrichment increases continuously. The value of the plateau, which is very
important to reach significant and accurate ^{15}N determination, is
effectively dependent on the rate of ^{15}N administration. The enrichments
of urinary urea in these two studies performed on preterm children are in
agreement although the rate of ^{15}N administration was 50 times higher in
our study.

In adult subjects (Steffee et al., 1976) where higher enrichments were
observed when taking into account the fact that a 10 times lower rate of
[15]N administration was given as compared with the present study (Fig. 2).
The value of the plateau being inversely related to the protein turnover
(equation 2), the higher enrichment obtained reflects a lower protein

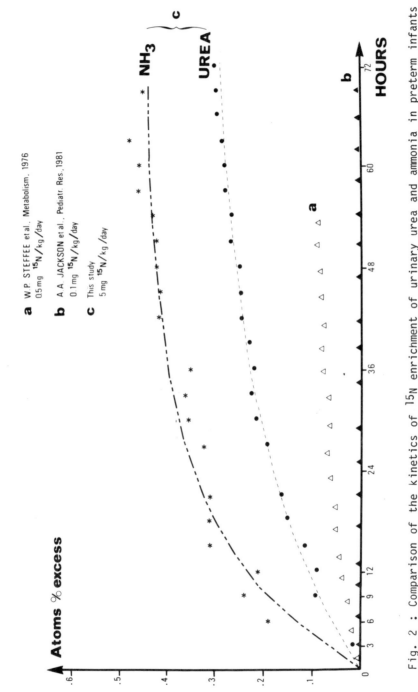

Fig. 2 : Comparison of the kinetics of ^{15}N enrichment of urinary urea and ammonia in preterm infants (2, 3) and in adult humans (1).

turnover in adult as compared with preterm children. The curves of urea and ammonia enrichment showed that glycine is preferentially transformed into ammonia (Waterlow et al., 1978). When protein synthesis is calculated from urea and ammonia plateau, differences as high as 60% can be observed. This discrepancy needs to be further studied.

Preliminary results of the first 4 VLBW infants are shown in Fig. 3. Whole body protein synthesis averaged 14 g/kg.d, whereas protein breakdown was 12 g/kg.d. Therefore, the mean protein gain was 2 g/kg.d, indicating that 7 times more protein was synthetised than laid down. The rate of protein synthesis found in the present study confirms recent results in VLBW fed a milk formula (Pencharz et al., 1981a).

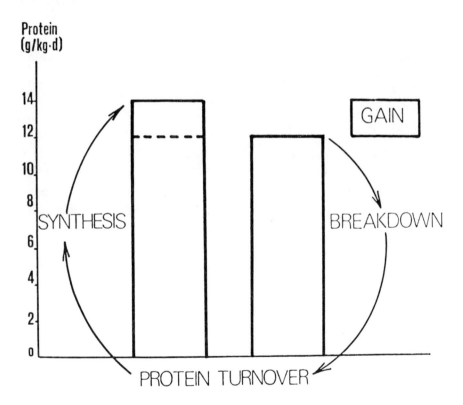

Fig. 3 : Rates of whole body protein synthesis and breakdown in very low birth weight infants (n = 4) fed 110 kcal/kg.d of metabolisable energy and 3.3 g prot/kg.d.

Since the synthesis of protein molecules entails a considerable amount of energy expenditure, it was of interest to examine the contribution of whole body protein synthesis to the overall resting energy expenditure during rapid growth. The theoretical cost of protein synthesis can be calculated from biochemical considerations, i.e. from the number of ATP molecules required to build a new protein molecule from individual amino acids. The cost incurred is due to the formation of peptide bonds between the amino acids. The cost of amino acid uptake and the synthesis of RNA has to be included (total : 6 mole ATP/mole amino acid incorporated into protein) (Flatt, 1978). If one takes an average value of 18 kcal. energy released per mole ATP used (Flatt, 1978), protein synthesis entails an energy expenditure of :

6 mole ATP/mole amino acid x 18 kcal/mole ATP = 108 kcal/mole amino acid

incorporated into protein, or approximately 1 kcal/g protein (one mole of an average amino acid is about 110 g).

Thus, the synthesis of 14 g/kg.d in a VLBW infant will require a theoretical energy expenditure of about 14 kcal/kg.d. The resting energy expenditure measured simultaneously was approximately, 60 kcal/kg.d. Hence, total body protein synthesis accounted for 23% of the energy expenditure. Probably at no other time in the postnatal period is the contribution of protein synthesis to energy expenditure of this magnitude. Calculations made by Pencharz et al. (1981b) based on literature data show that in 1-year-old infants, the contribution of protein synthesis to energy expenditure was about 10%, i.e. less than half of that observed in our VLBW infants.

CONCLUSION

Energy balance techniques combined with nitrogen balance measurements allowed both the study of how energy was utilised in VLBW infants and provided an estimate of lean and fat tissue accretions during the first few weeks of life. To explore the dynamic aspects of protein metabolism, estimates of rates of whole body protein synthesis and breakdown appear to be of great importance. The degree to which various dietary preparations can affect these variables in VLBW infants is a field worthy of intensive investigation.

ACKNOWLEDGEMENT

The help of J. Décombaz for bomb calorimetric measurements is greatly appreciated.

REFERENCES

Chessex, P., Reichman, B.L., Verellen, G.J.E., Putet, G., Smith, J.M., Heim, T. & Swyer, P.R. (1981) Influence of postnatal age, energy intake and weight gain on energy metabolism in the very low-birth weight infant. J. Pediat. 99:761-766.

Flatt, J.P. (1978) The biochemistry of energy expenditure. In : Recent Advances in Obesity Research, Vol. 2 (Ed. Bray, G) Newman Publishing, London, pp. 211-228.

Golden, M., Waterlow, J.C. & Picou, D. (1977) The relationship between dietary intake, weight change, nitrogen balance and protein turnover in man. Am. J. Clin. Nutr. 30:1345-1348.

Gudinchet, F., Schutz, Y., Micheli, J.L., Stettler, E. & Jéquier, E. (1982) Metabolic cost of growth in very low birth weight infants. Pediat. Res. 16:1025-1030.

Jackson, A.A., Shaw, J.C.L., Barber A. & Golden, M.H.N. (1981) Nitrogen metabolism in preterm infant fed human donor breast milk : the possible essentiality of glycine. Pediat. Res. 15:1454-1461.

Largo, R.H., Wälli, R., Duc, G., Fanconi, A. & Prader, A. (1980) Evaluation of perinatal growth. Helv. Paediatr. Acta 35:419-436.

Pencharz, P.B., Masson, M., Desgranges, F. & Papageorgiou, A. (1981a) Total-body protein turnover in human premature neonates : effects of birth weight, intra-uterine nutritional status and diet. Clin. Sci. 61:207-215.

Pencharz, P.B., Parsons, H., Motil, K. & Duffy, B. (1981b) Total body protein turnover and growth in children : is it a futile cycle ? Medical Hypotheses 7:155-160.

Picou, D. & Taylor-Roberts, T. (1969) The measurement of total protein synthesis and catabolism and nitrogen turnover in infants in different international states and receiving different amounts of dietary protein. Clin. Sci. 36:283-296.

Prader, A., Largo, R.H., Wälli, R. & Fanconi, A. (1980) Schweizerische Wachstumskurven von der 28. Schwangerchaftswoche bis zum 18. Lebensjar. Helv. Paediatr. Acta (Suppl. 45) p. 32 (abstract).

Schutz, Y. (1979) Estimates of the energy cost of growth in young children. Int. J. Vitam. Nutr. Res. (Suppl. 20) pp. 113-124.

Spady, D.W., Payne, P.R., Picou, D. & Waterlow, J.C. (1976) Energy balance during recovery from malnutrition. Am. J. Clin. Nutr. 29:1073-1088.

Steffee, W.P., Goldsmith, R.S., Pencharz, P.B., Scrimshaw, N.S. & Young, V.R. (1976) Dietary protein intake and dynamic aspects of whole body nitrogen metabolism in adult humans. Metabolism 25:281-297.

Thomson, J.F. (1963) Biological Effects of Deuterium. Pergamon Press, New York, pp. 1-130.

Waterlow, J.C., Garlick, P.J. & Millward, D.J. (1978) Measurement of protein turnover in the whole body with [15]N. In : Protein Turnover in Mammalian Tissues and in the Whole Body (Ed. Waterlow, J.C., Garlick, P.J.) North Holland Publishing Company, Amsterdam, New York, Oxford, pp. 252-299.

Widdowson, E.M. (1981) "Nutrition" in Scientific Foundations of Paediatrics, 2nd ed. (Ed. Davis, J.A., Dobbing, J.) William Heinemann Medical Books, London, p. 43.

ENERGY FUEL AND HORMONAL PROFILE IN EXPERIMENTAL OBESITIES

Bernard JEANRENAUD

Metabolic Research Laboratories, Faculty of Medicine,
University of Geneva, Geneva, Switzerland

SUMMARY

Several types of experimental obesities are characterized by the occurrence of an early hypersecretion of insulin that produces an increase in both triglyceride secretion by the liver and fat deposition in adipose tissue. This hypersecretion of insulin, together with other ill-defined factors, is subsequently responsible for a state of insulin resistance.

The early oversecretion of insulin in hypothalamic and genetic (e.g. fa/fa rats) obesities can be experimentally demonstrated. Thus, within 20 min of acute lesion of the ventromedial hypothalamus (VMH), glucose-induced insulin secretion is greater in lesioned than in non-lesioned control rats; this increase can be blocked by superimposed, acute vagotomy. Moreover, an infusion of glucose to 17-day-old, pre-weaned control and genetically pre-obese rats (i.e. animals genetically-determined to become obese but with a normal body weight at this age) elicits much greater insulinaemia in the pre-obese than in the controls, despite similar basal, pre-infusion values in both. This increased insulin secretion in the pre-obese rats can be restored to normal by pre-treating them acutely with the cholinergic inhibitor, atropine. Thus, in these two types of obesity, an increased vagal tone appears to be of importance for the early occurrence of insulin over-secretion. Hyperinsulinaemia produced by increased tone of the vagus nerve appears to be reinforced by the decreased activity of the sympathetic system observed in obese rodents.

In many obese rodents, plasma growth hormone levels are abnormally low. The inadequate secretion of this hyperglycaemic hormone may explain why, in some types of obesity syndrome, hyperglycaemia is not necessarily present, despite insulin resistance.

Insulin resistance in experimental obesities has been shown to occur at the level of the adipose tissue, the muscles and more recently, the liver. The latter has been demonstrated using the in vivo euglycaemic clamp technique; thus, glycogenolysis of genetically obese (fa/fa) rats could not be shut off, as in controls, by either basal or increased plasma insulin levels. This particular pathway is therefore insulin resistant.

The precise etiology of the early over-secretion of insulin in VMH-lesioned rats is, however, unknown : with VMH lesions, the origin is clearly the central nervous system (CNS), but the pathways actually interrupted by the lesions and those responsible for the hyperactivity of the vagus, remain to be determined. By analogy, it can be hypothesized that the cause of the early over-secretion of insulin in genetically obese rodents could also be in the CNS, though so far it has not been proven. This hypothesis is based on the observation that several CNS disorders arise in this type of obese rodent.

* * *

INSULIN HYPERSECRETION

In several types of rats or mice that become obese following lesioning of their hypothalami, or in rodents that spontaneously become obese (e.g. due to the presence of double recessive genes such as fa, ob, db, responsible for the syndrome), a common link (Fig. 1) is the occurrence of hyper-insulinaemia (Assimacopoulos-Jeannet & Jeanrenaud, 1976; Jeanrenaud, 1979). Hyperinsulinaemia overstimulates the hepatic parenchyma thereby increasing lipogenic pathways and resulting in : a) fatty livers (Jeanrenaud, 1978); b) increased hepatic very low density lipoprotein (VLDL) secretion (Karakash et al., 1977). Also, hyperinsulinaemia has been shown to overstimulate adipose tissue with the following consequences : a) increased in situ lipogenesis and/or b) augmented uptake of circulating VLDL due to the likely stimula-tion, by increased plasma insulin levels, of activity of the insulin-sensitive lipoprotein lipase (LPL), an enzyme that is required for VLDL-triglyceride uptake (Gruen et al., 1978). In our mind, these abnormalities are, at least in the above-mentioned rodents, some of the main reasons for obesity. This is not to claim that hyperphagia has no relevance. Indeed, it increases the amount of incoming nutrients, further stimulating insulin secretion; it can, therefore, be viewed as an aggravation factor (Martin et al., 1974). Much data are available, however, to show that obesity can occur in the absence of hyperphagia and that, under those conditions, increased plasma insulin levels are major determinants of obesity (Bernardis et al., 1975; Hustvedt et al., 1976). One should mention that in several species, in the genetically obese (fa/fa) rats in particular, plasma growth hormone levels are decreased as well. As this hormone is long-acting and lipolytic, its absence would, of course, further favour obesity (Martin & Jeanrenaud, 1983).

INSULIN RESISTANCE - A MULTIFACTORIAL PATHOLOGY

The evolution of most obese rodents is towards a state of insulin resis-tance that may have varying characteristics : a) normaglycaemia at the cost of excessive insulin secretion; b) hyperglycaemia with normal glucose

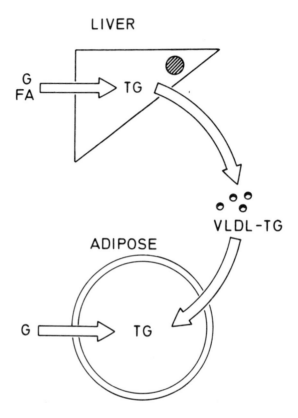

Fig. 1 : Consequences of hyperinsulinemia on lipogenic pathways of obese rodents.

Many experiments have shown that hyperinsulinaemia of obese rodents results in an overstimulation of pathways involved in lipid synthesis and storage. This overstimulation is schematized by the heavy arrows. In the liver, hyperinsulinaemia produces increased lipogenesis (TG = triglycerides) from various substrates (e.g. from fatty acids, FA; glucose, G, etc.), increased synthesis of very low density lipoprotein (VLDL-TG), with resulting increased hepatic VLDL-TG secretion and fatty livers (hatched circle).

In the adipose tissue, hyperinsulinaemia has two main consequences : a) increased in situ lipogenesis from various substrates (e.g. glucose, G); b) increased uptake of circulating VLDL-TG due to increased activity of lipoprotein lipase activity, an insulin-responsive enzyme necessary for actual uptake of VLDL-TG. The overall result of these two main consequences is an increased fat storage within each adipocyte, the size of which increases, hence occurrence of obesity. (Based on Assimacopoulos-Jeannet & Jeanrenaud, 1976; Karash et al., 1977; Gruen et al., 1978; Jeanrenaud, 1978).

tolerance together with hyperinsulinaemia; c) hyperglycaemia with abnormal glucose tolerance and hyperinsulinaemia. The insulin resistance is thus a multifactorial pathology as summarized in Fig. 2. In all insulin-sensitive tissues, a decreased insulin receptor number has been observed that is due to prevailing hyperinsulinemia (Le Marchand-Brustel & Freychet, 1978; Czech & Reaven, 1978). Indeed, the hormone down-regulates its own insulin receptor, i.e. the more insulin, the less receptor and conversely (Kahn, 1978). Decreased receptor number is responsible for decreasing the sensitivity of the tissues to the hormone, i.e. more hormone is needed to produce any given, insulin-sensitive metabolic effect in normal tissues (Kahn, 1978). Other abnormalities may be triggered by hyperinsulinaemia but a clear-cut cause-effect relationship is more difficult to prove than for decreased insulin receptor number. These other defects can be found (in muscle, adipose tissues) to be located at many different levels and can be summarized as follows : a) decreased transport capacity of the plasma membrane glucose transport systems (Le Marchand-Brustel et al., 1978; Zaninetti et al., 1983); b) intracellular abnormalities that prevent normal utilization of phosphorylated glucose intermediates leading, for example, to a decreased muscle glycolysis and/or decreased glycogen synthesis (Czech et al., 1978; Crettaz et al., 1980) together with a possible channelling of substrates into fat and hence, increased muscle triglyceride content. In fat cells, analogous intracellular defects prevent fatty acid synthesis and the enlarged fat cells of obese animals probably remain enlarged despite decreased de novo fat synthesis. Indeed LPL activity and esterification capacity of those cells remain high enough to permit VLDL uptake (Gruen et al., 1978; Cushman et al., 1978).

INSULIN RESISTANCE IN LIVER

Until recently, insulin resistance in the liver has not been studied, since, for unknown reasons, normal isolated perfused livers or isolated hepatocytes even when perfectly preserved, responded poorly to insulin when hepatic glucose production was investigated. As it was not possible to measure the expected insulin-dependent decrease in hepatic glucose production

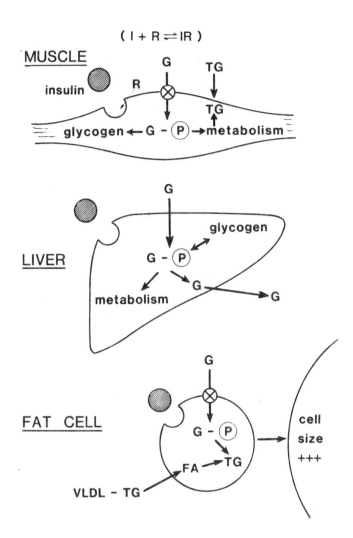

Fig. 2 : Effects of hyperinsulinaemia (together with other unknown alterations) on the occurrence of insulin resistance, hence on that of inappropriate glucose homeostasis in obese rodents. Insulin (I) normally binds to specific target tissue receptors (R) to form IR complexes that are ultimately responsible for the final metabolic responses. Hyperinsulinaemia (shown here by hatched circles) is responsible for the occurrence of a decrease in the number of receptors at the plasma membranes of target tissues (exaggerated down-regulation of R by I). Target tissues, therefore, become less insulin-sensitive.

Hyperinsulinaemia together with other ill-defined factors are also responsible for the following alterations that further aggravate insulin resistance :

Muscle and adipose tissue : Defective glucose transport (shown by ⊗), i.e. the translocation of circulating blood glucose into these cells, is slowed down or does not occur any more.

Muscle : Intracellular phosphorylated glucose (G-P) is poorly converted to glycogen and poorly metabolized, because of many intracellular abnormalities. Intracellular phosphorylated glucose metabolism is partly channelled to triglyceride synthesis (TG) : increased TG content in muscles of obese rodents is frequently encountered (possibly also due to increased lipoprotein lipase activity, an enzyme responsible for TG uptake). Increased TG content may play a role in abnormal utilization of glucose and abnormal glycogen synthesis of muscles from obese rodents.

Liver : Intracellular phosphorylated glucose (GP) is abnormally handled. These abnormalities are still poorly substantiated. It has been clearly shown, however, that hepatic glucose production is no longer adequately shut off by insulin in some types of genetically obese rodents.

Adipose tissue : Intracellular phosphorylated glucose (GP) is poorly converted to triglyceride (TG) because of many intracellular abnormalities. The enlarged fat cells of obese rodents probably remain enlarged, despite decreased TG synthesis, because lipoprotein lipase does not easily become insulin resistant, very low density lipoprotein (VLDL-TG) uptake remaining therefore increased, as mentioned in Fig. 1.

(Based on Assimacopoulos- Jeannet & Jeanrenaud, 1976; Czech & Reaven, 1978; Czech et al., 1978; De Fronzo, 1978; Kahn, 1978; Le Marchand-Brustel & Freychet, 1978; Le Marchand- Brustel et al., 1978; Crettaz et al., 1980; Terrettaz & Jeanrenaud, 1983; Zaninetti et al., 1983).

in normal rodents, a resistance to this process was a fortiori impossible to test in obese models. To circumvent these difficulties, we decided to apply to small rodents the hyper- and euglycemic clamp methods known to work satisfactorily in humans (De Fronzo, 1978). This technique is described elsewhere (Karakash et al., 1977). The main advantage in this approach lies in the ability of measuring in vivo hepatic glucose production. The experiments carried out in lean and obese Zucker (fa/fa) rats gave the following results. In lean rats, when insulinaemia was raised from normal (2 ng/ml) basal to a new higher (about 7 ng/ml) steady state level, hepatic glucose production was shut off. In contrast, in genetically obese fa/fa rats hepatic glucose production was, under basal conditions, as high as in normal controls despite the fact that the obese were hyperinsulinaemic, indicating that the hepatic glucose production was not inhibited even by the high (10 ng/ml at least) basal plasma insulin levels, and that the process was, therefore, insulin-resistant. Moreover, when hyperinsulinaemia of obese rats was further increased to a new higher steady state level of 20 ng/ml, hepatic glucose production was not shut off (Terrettaz & Jeanrenaud, 1983).

The underlying mechanisms responsible for this newly described hepatic insulin resistance are unknown but the enzymes responsible for hepatic glucose handling are currently under investigation.

CNS-ENDOCRINE PANCREAS AXIS

The cause of hyperinsulinaemia of obese rodents remains unknown (Rohner-Jeanrenaud et al., 1983a). Since (a) hyperphagia per se cannot be held responsible for the occurrence of obesity, hyperinsulinaemia being a more likely cause of the syndrome, (b) "peripheral" hyperinsulinaemia-linked alterations measured in liver, adipose tissue and muscles of genetically obese rodents or in rodents made obese by lesions of the hypothalamus are analogous, and finally (c) central nervous system (CNS) abnormalities are numerous in both genetic and hypothalamic obesity, it was thought that a functional CNS-endocrine pancreas axis might exist in normal rats. This axis might be abnormally regulated in the various obesity syndromes mentioned, thereby causing hyperinsulinemia. Various CNS manipulations have therefore been carried out in an attempt to show that such an axis may exist.

It has been found by several authors that the CNS normally exerts an overall tonic inhibitory influence on the endocrine pancreas that ventro-medial hypothalamic (VMH) lesions somehow alleviate (Rohner-Jeanrenaud et al., 1983a). Thus, when normal anaesthetized rats are lesioned in the VMH, glucose-induced insulin secretion becomes higher than that of controls, and this within minutes following the lesions. This rapidly occurring over-secretion of insulin is mediated by the vagus nerve as it is completely and immediately reversed by acute vagotomy (Berthoud & Jeanrenaud, 1979). When acute vagotomy is not performed, hyperinsulinaemia produced by VMH lesions persists and increases with time even in the absence of hyperphagia (Rohner-Jeanrenaud & Jeanrenaud, 1980). The increased activity of the cholinergic system that follows VMH lesions is also substantiated in several in vitro experiments in which, as shown in Fig. 3, the increase in insulin secretion observed in pancreases from VMH-lesioned rats is restored to normal by infusion of atropine (Rohner-Jeanrenaud & Jeanrenaud, 1981).

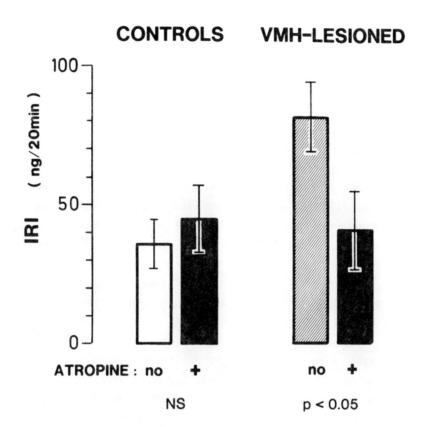

KRB buffer , 10 mM arginine , 0 glucose , 25 µM atropine , perfusion time : 20 min .

Mean ± SEM (n = 4 – 6)

Fig. 3 : Effect of ventromedial hypothalamic (VMH) lesions on the in vitro insulin secretion by subsequently perfused pancreas : effect of atropine. Perfusions were carried out with Krebs-Ringer bicarbonate buffer with 10 mM arginine. (Based on Rohner-Jeanrenaud & Jeanrenaud, 1981).

The role of the vagus nerve in being partly responsible for the spontaneous genetically-linked hyperinsulinaemia deserves some mention as this syndrome represents a true pathological situation. It is observed that in 17-day-old pre-obese Zucker rats (i.e. in preweaned rats that will become obese but which, at the experimental time, have a normal body weight and normal basal insulinaemia) an i-v glucose challenge produces rises in plasma insulin levels that are much higher than that of the respective controls. More importantly, this increased insulin secretion of pre-obese rats is abolished by cholinergic blockade, clearly indicating a parasympathetic origin of insulin oversecretion. Moreover, when genetically obese animals are adult, the arginine or glucose-induced secretion of insulin is much higher than normal and these abnormalities can be corrected by pharma-cological or surgical blockade of the vagus nerve (Rohner-Jeanrenaud et al., 1983b). Since genetically obese rats have several CNS abnormalities, as mentioned above, these data suggest that genetically linked CNS dys-functions, possibly and partly hypothalamus-related would alter the fine balance that normally tunes the respective parasympathetic and sympathetic outflow, resulting in an overactivity of the vagus nerve. This would be responsible for the occurrence of hyperinsulinaemia, obesity, and ultimately, insulin resistance. Thus, hypothalamic (Fig. 3) and genetically-linked (Table I) hyperinsulinaemia may have analogous underlying etiologies (Rohner-Jeanrenaud et al., 1983a).

Increased vagal nerve activity due to CNS disturbances, together with other unknown "trophic" factors, may explain why most hyperinsulinaemic animals eventually have increased B-cell mass, as depicted in Fig. 4 (Han & Frohman, 1970; Inoue et al., 1978). Such increased B-cell mass may become an important pathological trait since incoming nutrients (e.g. glucose, amino acids) are continuously prone to trigger oversecretion of insulin by the mere existence of such increased B-cell mass. This is in keeping with the finding that to best prevent long-term consequences of VMH lesions, vagotomy has to occur prior to the VMH lesions (Cox & Powley, 1981).

The use of VMH lesions has permitted the conclusion that CNS modulation of the endocrine pancreas was mediated not only by modulation of the vagus tone but by that of the sympathetic one as well, as shown in Fig. 5. Indeed, the overall sympathetic nervous tone appears to be decreased following VMH lesions (Inoue & Bray, 1979; Bray et al., 1981). When VMH-lesioned animals are placed under stressful situations which elicit a sympathetic response, the latter is blunted. The sympathetic nerve-mediated increase in plasma free fatty acids (FFA) that occurs in normal rats following induced cellular glucopenia is reduced in VMH-lesioned animals, as is the sympathetic response following exercise, fasting or cold exposure (Nishizawa & Bray, 1978). The observed increase in basal or glucose-induced insulin secretion of VMH-lesioned rats can be reduced not only by atropine but by epinephrine as well, the combination of the two drugs bringing insulinaemia back to normal values (Inoue & Bray, 1980). This suggests that following VMH lesions, increased vagal and decreased sympathetic tones are of importance in bringing about hyperinsulinaemia and its metabolic consequences.

Table I : Comparison between the changes in plasma insulin and glucose levels of acutely VHM-lesioned, pre-obese and genetically obese (fa/fa) rats

	VMH lesioned	Pre-obese and obese (fa/fa) rats
1. Basal glycemia	normal	normal
2. Basal insulinemia or basal insulin secretion	normal	normal
3. Substrate-induced induced secretion	increased	increased
	normalized by cholinergic blockade	normalized by cholinergic blockade
	normalized by acute vagotomy	ameliorated by acute vagotomy

From Berthoud & Jeanrenaud (1979); Rohner-Jeanrenaud & Jeanrenaud (1980); Rohner-Jeanrenaud et al. (1983a, 1983b).

Acute VMH lesion   7day VMH lesion

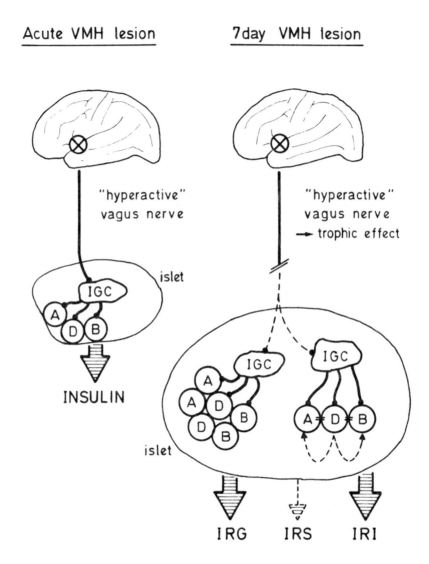

Fig. 4 : Acute lesions of the ventromedial hypothalamus (VMH) result in a hyperactivity of the vagus nerve, hence hypersecretion of insulin. In chronically VMH lesioned rats hyperactivity of the vagus nerve persists and is accompanied (due to unknown trophic effects) by increased size of the islets of Langerhans. IRI = immunoreactive insulin; IRS = immunoreactive somatostatin; IRG = immunoreactive glucagon. (Based on Han & Frohman, 1970; Inoue et al., 1978; Berthoud & Jeanrenaud, 1979; Rohner-Jeanrenaud & Jeanrenaud, 1980; 1981).

One should emphasize that VMH lesions do not give to this particular site the authority of being a "center"; one can only state that the interruption of this particular circuitry acts on CNS homeostasis in such a way that both the vagal and the sympathetic outflows are altered and that an analogous dysfunction may conceivably prevail in genetically obese rodents, as schematized in Fig. 5 (Rohner-Jeanrenaud et al., 1983a).

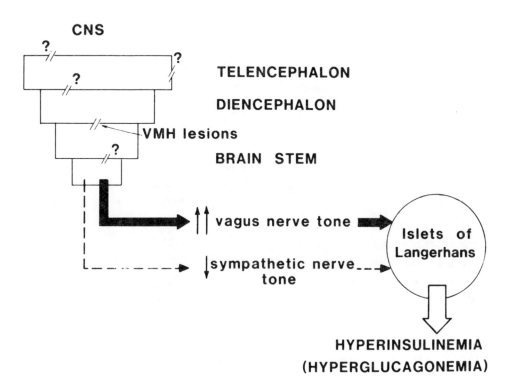

Fig. 5 : CNS-endocrine pancreas axis : ventromedial hypothalamic (VMH) lesions.

Consequences of the interruption of the CNS circuitry at the level of the ventromedial hypothalamus (VMH), upon the parasympathetic and sympathetic tones and, ultimately, upon insulin secretion. Question-marks indicate that VMH lesions alter the normal CNS homeostasis not only at the VMH per se but at (largely unknown) related CNS sites : This is to stress that the VMH should not necessarily be considered as "a center" but rather as an important site of convergent information.

(Based on Berthoud & Jeanrenaud, 1979; Bray et al., 1981; Inoue & Bray, 1979; 1980; Nishizawa & Bray, 1978; Rohner-Jeanrenaud, 1980; 1981; Rohner-Jeanrenaud et al., 1983a).

Most studies carried out to determine the respective role of the vagus or the splanchnic nerve on endocrine pancreas secretions have been achieved by stimulations of efferent pathways (Woods & Porte, 1978; Miller, 1981). One should emphasize, however, that 75-90% of all fibers in the abdominal vagus and 50% of all splanchnic fibers are afferent fibers that bring information to the CNS, information that is somehow processed there and followed by meaningful efferent responses (Kral, 1981). For example, glucose "sensors" of the small intestine are also functionally related to vagal fibers (Mei, 1978; Mei et al., 1980). Several types of "sensors" exist in the region of the portal vein and within the liver that send vagally-mediated afferences (Russek, 1963; Carobi & Magni, 1981) in particular to the hypothalamus (Schmitt, 1973). Among these hepatic, vagally-mediated "sensors", some are glucose responsive (Sawchenko & Friedman, 1979).

An illustration of a potential regulation of insulin secretion that would be triggered by peripheral glucose "sensors" is provided by a series of electrophysiological investigations (Niijima, 1979) : a) hyperglucaemia within the portal vein produces a vagus nerve mediated signal that results in decreased adrenal nerve efferent activities and increased vagal efferent activities that eventually reach the pancreas. This would indicate, upon extrapolation, that a signal from the periphery (e.g. the liver) elicited by ample availability of glucose would change CNS-integrated responses, which would be compatible with a decreased adrenaline output and an increased insulin secretion; b) conversely, low glycaemia within the portal vein elicits another vagus nerve mediated signal that produces increased adrenal nerve efferent activity and decreased vagal efferents. This would be compatible with changes in hormonal output such as, respectively, increased adrenaline, and decreased insulin secretions.

With regard to afferent signals to the brain and efferent responses, the cephalic phase insulin secretion merits mention. It has been proposed that the responses occurring at the pre-absorptive stage modify metabolic functions in the same way as the nutrients do when they ultimately reach the internal milieu. They are therefore referred to as anticipatory reflexes that seem to "prepare" the release of those factors that allow metabolization of the ingested substances, so that these substances are best utilized (Nicolaïdis, 1978).

Early reflex phase insulin secretion is a part of these anticipatory reflexes and may be defined as an autonomic-endocrine reflex that is triggered by the sight/smell of food or by the contact (not by the absorbtive consequences) of incoming nutrients with various sensor cells of the oropharynx and the stomach (Rohner-Jeanrenaud et al., 1983a). Early reflex phase insulin secretion consists in a rapid (within 2-3 min), transient (disappearance within 5-9 min), small (delta over baselines of the order of 3 ng/mg) increase in plasma insulin levels that occurs without any change in glycaemia (Berthoud & Jeanrenaud, 1982). Further, it comprises an afferent sensory neural limb, a CNS integration circuitry that remains to be defined in more detail but which includes the hypothalamus and an efferent limb which is the vagus nerve and which produces the type of insulin secretion just mentioned (Rohner-Jeanrenaud et al., 1983a).

Early cephalic phase insulin secretion may have an important physiological relevance, that may be coined "priming" or "preparing" the endocrine pancreas. This "priming" effect is not understood mechanistically but can be demonstrated as follows : When early cephalic phase insulin secretion is present, meal-induced insulin secretion displays a well-characterized first and second phase secretory pattern; hyperglycaemia is moderate and then returns to basal values. In contrast, when early phase insulin secretion is absent (denervation of the pancreas), the subsequent insulin secretion does not retain a biphasic pattern. The first peak insulin secretion is missing, the subsequent rise in plasma insulin levels is delayed ultimately to reach values that are higher than those seen in animals showing the early phase insulin secretion (Steffens, 1981; Strubbe & van Wachem, 1981).

To sum up, early cephalic phase insulin secretion appears to minimize, via yet unknown mechanisms, the subsequent insulin requirement for maintaining normal glucose tolerance and may therefore be implicated in the occurrence of hyperinsulinaemia of various types of obesity. Thus, lack of or abnormal early reflex insulin secretion could be responsible for a greater peripheral hyperglycaemia that would in turn overstimulate the B-cells. Lack of or abnormal early phase insulin secretion might thus be responsible, due to concomitant excessive meal-induced hyperinsulinaemia and hyperglycaemia, for the reported increased channelling of glucose into fat

and away from the maintenance of lean body mass that occurs in obesity
(Steffens, 1981).

The CNS-endocrine pancreas axis may conceivably comprise humoral factors
as well. Data from our laboratories have shown that electrical stimulation
of the ventrolateral hypothalami (VLH) of anaesthetized rats under phentol-
amine (α-adrenergic blockade) produced a marked increase in plasma insulin
levels which could not be prevented by superimposed vagotomy, chordotomy or
pharmacological blockade of the β-adrenergic receptors (Berthoud et al.,
1980). This suggests that a humoral factor had been released during VLH
stimulation, a factor(s) that would promote insulin secretion.

Further investigations indeed showed the presence in the VLH of normal
rats, of a factor(s) capable of increasing insulin secretion when injected
in vivo to recipient rats (Bobbioni & Jeanrenaud, 1982) or in vitro when
infused into normal perfused pancreases (Bobbioni & Jeanrenaud, 1983). It is
worth mentioning that the effect of this factor(s) was shown not to be
related to the possible presence (in VLH extracts) of noradrenaline, acetylcholine, enkephalins, gastrin or cholecystokinin (CCK). The polypeptidic
nature of this hypothalamic factor(s) is suggested by its sensitivity to
some proteolytic enzymes. Upon a first purification (Sephadex G 50 chromatography) the insulin secretion promoting activity of VLH corresponds to
small molecular weight components. When infused alone into perfused
pancreases (at a final 27 µg/ml protein concentration), the active
fraction of VLH obtained from this first purification elicited a 12 ng/ml
peak insulin output. Furthermore, it enhanced both the first and second
peaks of glucose- or amino acid-induced insulin secretion. A second step
purification (Biogel) of VLH extracts revealed that the molecular weight of
the factor(s) was 1200 or smaller and that it was endowed with a powerful
stimulatory effect on insulin secretion since, when added alone to perfused
pancreases (at a 2 µg protein final concentration), a 15 ng/5 min.
increase in insulin output was obtained (Bobbioni & Jeanrenaud, 1982; 1983).

For this and all other analogous factors reported to have physiological
relevance, it remains to be established whether they might act as neurotransmitters and stimulate fibers that are facilitating insulin secretion,

whether they might be neurally transported and act as neuromodulators at the endocrine pancreas or whether they might reach the B-cells following their actual release into the blood (Rohner-Jeanrenaud et al., 1983a). It is encouraging to note that factors analogous to those summarized above (Bobbioni & Jeanrenaud, 1982) appear to be increased in hypothalamus in, so far, two types of obesities, the hypothalamic and the genetic (fa/fa) and may possibly play a rôle in the occurrence of these syndromes.

REFERENCES

Assimacopoulos-Jeannet, F. & Jeanrenaud, B. (1976) The hormonal and metabolic basis of experimental obesity. Clinics in Endocrinol. Metab. 5(2): 337-365.

Bernardis, L.L., Goldman, J.K., Chlouverakis, C. & Frohman, L.A. (1975) Six-month follow-up in weanling rats with ventromedial and dorsomedial hypothalamic lesions : Somatic, endocrine, and metabolic changes. J. Neurosci. Res. 1:95-108.

Berthoud, H.R. & Jeanrenaud, B. (1979) Acute hyperinsulinemia and its reversal by vagotomy after lesions of the ventromedial hypothalamus in anaesthetized rats. Endocrinology 105:146-151.

Berthoud, H.R., Bereiter, D.A. & Jeanrenaud, B. (1980) Rôle of the autonomic nervous system in the mediation of LHA electrical stimulation-induced effects on insulinaemia and glycaemia. J. Auton. Nerv. Sys. 2:183-198.

Berthoud, H.R. & Jeanrenaud, B. (1982) Sham feeding-induced cephalic phase insulin release in the rat. Am. J. Physiol. 252:E280-E285.

Bobbioni, E. & Jeanrenaud, B. (1982) Effect of rat hypothalamic extract administration on insulin secretion in vivo. Endocrinology 110:631-636.

Bobbioni, E. & Jeanrenaud, B. (1983) A partially purified rat hypothalamic extract stimulates insulin secretion in vitro (submitted for publication).

Bray, G.A., Inoue, S. & Nishizawa, Y. (1981) Hypothalamic obesity. The autonomic hypothesis and the lateral hypothalamus. Diabetologia 20:366-377.

Carobi, C. & Magni, F. (1981) The afferent innervation of the liver : a horseradish peroxidase study in the rat. Neurosci. Lett. 23:269-274.

Cox, J.E. & Powley, T.L. (1981) Prior vagotomy blocks VMH obesity in pair-fed rats. Am. J. Physiol. 240:E573-E583.

Crettaz, M., Prentki, M., Zaninetti, D. & Jeanrenaud, B. (1980) Insulin resistance in soleus muscle from obese Zucker rats. Involvement of several defective sites. Biochem. J. 186:525-534.

Cushman, S.W., Zarnowski, M.J., Franzusoff, A.J. & Salans, L.B. (1978) Alterations in glucose metabolism and its stimulation by insulin in isolated adipose cells during the development of genetic obesity in the Zucker fatty rat. Metabolism 27:1980-1940.

Czech, M.P. & Reaven, G.M. (guest editors) (1978) Insulin insensitivity. Metabolism 27 (Suppl. 2) pp. 1829-2014.

Czech, M.P., Richardson, D.K., Becker, S.G., Walters, L.G., Gitomer, W. & Heinrich, J. (1978) Insulin response in skeletal muscle and fat cells of the genetically obese Zucker rat. Metabolism 27:1967-1981.

DeFronzo, R.A. (1978) Pathogenesis of glucose intolerance in uraemia. Metabolism 27:1866-1880.

Gruen, R., Hietanen, E. & Grennwood, M.R.C. (1978) Increased adipose tissue lipoprotein lipase activity during the development of the genetically obese rat (fa/fa). Metabolism 27:1955-1966.

Han, P.W. & Frohman, L.A. (1970) Hyperinsulinaemia in tube-fed hypophy-sectomized rats bearing hypothalamic lesions. Am. J. Physiol. 219:1632-1636.

Hustvedt, B.E., Lovo, A. & Reichl, D. (1976) The effect of ventromedial hypothalamic lesions on metabolism and insulin secretion in rats on a controlled feeding regimen. Nutr. Metab. 20:264-271.

Inoue, S. & Bray, G.A. (1979) An autonomic hypothesis for hypothalamic obesity. Life Sci. 25:561-566.

Inoue, S. & Bray, G.A. (1980) Rôle of the autonomic nervous system in the development of ventromedial hypothalamic obesity. Brain Res. Bull. 5:119-125.

Inoue, S., Bray, G.A. & Mullen, Y.S. (1978) Transplantation of pancreatic β-cells prevents development of hypothalamic obesity in rats. Am. J. Physiol. 235:E266-E271.

Jeanrenaud, B. (1978) Hyperinsulinaemia in obesity syndromes : its metabolic consequences and possible etiology. Metabolism 27:1881-1892.

Jeanrenaud, B. (1979) Insulin and obesity. Diabetologia 17:133-138.

Kahn, C.R. (1978) Insulin resistance, insulin insensitivity, and insulin unresponsiveness : a necessary distinction. Metabolism 27:1893-1902.

Karakash, C., Hustvedt, B.E., Lovo, A., Le Marchand, Y. & Jeanrenaud, B. (1977) Consequences of ventromedial hypothalamic lesions on metabolism of perfused rat liver. Am. J. Physiol. 232:E286-E293.

Kral, J.G. (1981) Vagal mechanisms in appetite regulation. Intern. J. Obesity 5:481-489.

Le Marchand-Brustel, Y. & Freychet, P. (1978) Studies of insulin insensitivity in soleus muscles of obese mice. Metabolism 27:1982-1993.

Le Marchand-Brustel, Y., Jeanrenaud, B. & Freychet, P. (1978) Insulin binding and effects in the isolated soleus muscle of lean and obese mice. Am. J. Physiol. 234:E348-E358.

Martin, R.J. & Jeanrenaud, B. (1983) Growth hormone in obesity and diabetes : inappropriate hypothalamic control of secretion (submitted for publication).

Martin, J.M., Konijnendijk, W. & Bouman, P.R. (1974) Insulin and growth hormone secretion in rats with ventromedial hypothalamic lesions maintained on restricted food intake. Diabetes 23:203-208.

Mei, N. (1978) Vagal glucoreceptors in the small intestine of the cat. J. Physiol. 282:485-506.

Mei, N., Jeanningros, R. & Boyer, A. (1980) Rôle des glucorécepteurs vagaux de l'intestin dans la régulation de l'insulinémie. Reprod. Nutr. Dévelop. 20:1621-1624.

Miller, R.E. (1981) Pancreatic neuroendocrinology : peripheral neural mechanisms in the regulation of the islets of Langerhans. Endocrine Rev. 2:471-494.

Nicolaïdis, S. (1978) Rôle des réflexes antipateurs oro-végétatifs dans la régulation hydrominérale et énergétique. J. Physiol. (Paris) 74:1-19.

Niijima, A. (1979) Control of liver function and neuroendocrine regulation of blood glucose levels. In : Integrative Functions of the Autonomic Nervous System (Ed. Brooks, C. McC., Koizuni, K., Sato, A.) Elsevier/North-Holland Biomed. Press, pp. 68-83.

Nishizawa, Y. & Bray, G.A. (1978) Ventromedial hypothalamic lesions and the mobilization of fatty acids. J. Clin. Invest. 61:714-721.

Rohner-Jeanrenaud, F. & Jeanrenaud, B. (1980) Consequences of ventromedial hypothalamic lesions upon insulin and glucagon secretion by subsequently isolated perfused pancreases in the rat. J. Clin. Invest. 65:902-910.

Rohner-Jeanrenaud, F. & Jeanrenaud, B. (1981) Possible involvement of the cholinergic system in hormonal secretion by the perfused pancreas from ventromedial-hypothalamic lesioned rats. Diabetologia 20:217-222.

Rohner-Jeanrenaud, F., Bobbioni, E., Ionescu, E., Sauter, J.F. & Jeanrenaud, B. (1983a) CNS regulation of insulin secretion. Advances in Endocrine Disorders. Chap. X, Academic Press, New York (in press).

Rohner-Jeanrenaud, F., Hochstrasser, A.C. & Jeanrenaud, B. (1983b) Hyperinsulinaemia of pre-obese and obese fa/fa rats is partly vagus nerve mediated. Am. J. Physiol. (in press).

Russek, M. (1963) Participation of hepatic glucoreceptors in the control of intake of food. Nature 197:79-80.

Sawchenko, P.E. & Friedman, M.I. (1979) Sensory functions of the liver - a review. Am. J. Physiol. 236:R5-R20.

Schmitt, M. (1973) Influences of hepatic portal receptors on hypothalamic feeding and satiety centers. Am. J. Physiol. 225:1089-1095.

Steffens, A.B. (1981) The regulatory rôle of the central nervous system on insulin and glucagon release during food intake in the rat. In : Hormones and Cell Regulation 5 (Ed. Dumont J.E., Nunez, J.) Elsevier/North-Holland Biomed. Press, pp. 185-196.

Strubbe, J.H. & van Wachem, P. (1981) Insulin secretion by the transplanted neonatal pancreas during food intake in fasted and fed rats. Diabetologia 20:228-236.

Terrettaz, J. & Jeanrenaud, B. (1983) In vivo hepatic and peripheral insulin resistance in genetically obese (fa/fa) rats. Endocrinology (in press).

Woods, S.C. & Porte, D. Jr (1978) The central nervous system, pancreatic hormones, feeding, and obesity. In : Advances in Metabolic Disorders (Ed. Levine, R., Luft, R.), vol. 9, Academic Press, New York, San Francisco, London, pp. 283-312.

Zaninetti, D., Crettaz, M. & Jeanrenaud, B. (1983) Dysregulation of glucose transport in hearts of genetically obese fa/fa rats (submitted for publication).

NUTRIENT INTAKE AND ENERGY REGULATION IN PHYSICAL EXERCISE

Hans HOWALD* and Jacques DECOMBAZ

*Research Institute of the Swiss School for Physical Education
and Sports, 2532 Magglingen, Switzerland, and

Nestlé Products Technical Assistance Co. Ltd., Research Department,
1814 La Tour-de-Peilz, Switzerland

SUMMARY

Rates of energy expenditure as well as total daily energy
cost can be considerable during periods of exercise. In trained
athletes, expenditure can be as high as 380 kJ/min during short-
term maximal exercise. Training programmes of several hours'
duration lead to a daily nutrient intake of 25-35 MJ in most
Olympic sports.

The mobilization of the energetic fuels of the body is modu-
lated by the nature of the exercise. ATP and creatine phosphate
stores in muscle cells are depleted within seconds during maximal work. Glycogen is the main fuel for heavy exercise of a few
minutes' duration where performance capacity is limited by the
degree of lactate accumulation and intracellular acidosis. Oxidation of both glucose and free fatty acids supplies the energy
needed for exercise lasting more than two minutes, the relative
contribution of lipids increasing with a longer duration or a
lower intensity of the muscular work. Intramuscular stores of
glycogen and triglycerides may be almost completely depleted in

long-lasting exercise, e.g. a 100 km run. Under these conditions, glycogen stores in the liver and triglycerides in adipose
tissue contribute approximately 70% of the energy need whereas
5-10% of the supply comes from oxidation of amino acids.

Although adequate nutrition for exercise could be achieved
through the intake of a well-balanced diet, the regulation of
energy utilization can be influenced by the sources of food
energy, by dietary modifications before exercise or by nutrient
supplements during exercise.

Intake before exercise of fructose or medium-chain triglycerides, both only weakly insulinogenic compared to glucose,
leads to changes in blood substrates and metabolites. However,
neither glycogen depletion in the working muscles nor performance capacity was influenced by a single meal containing
this particular carbohydrate or lipid.

Mobilization of free fatty acids in adipose tissue can be
enhanced by caffeine or depressed by nicotinic acid. Since the
rate of free fatty acid oxidation in skeletal muscle depends on
the blood concentration of this substrate, energy regulation
during exercise and work output are considerably influenced by
the ingestion of such substances.

* * *

ENERGY EXPENDITURE AND MUSCLE METABOLISM

Physical activity is one of the most important factors in the calculation
of daily energy needs in man. Athletes are extreme cases not only with
respect to their sporting performances, but also as far as nutritional
aspects are concerned. In most of the Olympic sports, training means a daily
energy expenditure of 25 to 35 MJ (Nöcker, 1974), and thus the nutritional
needs of some athletes exceed by far those of any professional group in our
modern society.

Total energy output depends on the type, the intensity and the duration of a given exercise. In well-trained athletes, it can be as high as 380 kJ/min. in weight lifting, 210 kJ/min. in 400 meter sprinting, 150 kJ/min. in rowing and 100 kJ in marathon running. The decrease in energy output with the duration of exercise and the metabolic pathways supplying the energy for muscular work are summarized in Fig. 1.

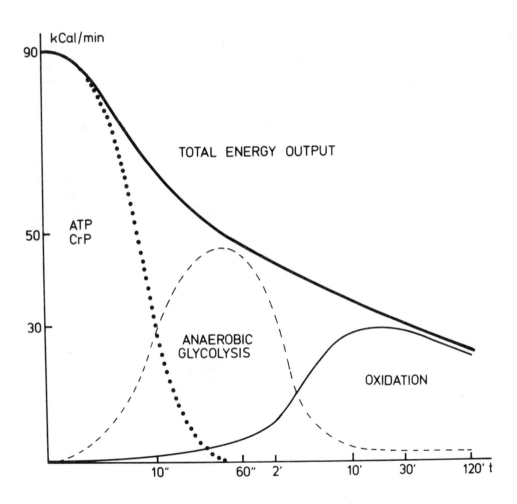

Fig. 1 : Time-dependent reduction in maximum energy output and sequence of metabolic pathways supplying energy to the working skeletal muscle. Time is given on a logarithmic scale so that very short bouts of exercise can be compared with marathon running.

The limiting factors in short-term exercise lasting no more than 15 seconds (weight lifting, 100 meter run etc.) are muscle strength and availability of creatine phosphate (CrP). Performance capacity in events lasting 1 to 2 minutes (400-800 meter races) is limited by the efficiency of anaerobic glycogenolysis and by the ability to tolerate blood lactate concentrations exceeding 20 mmoles/l. As soon as the duration of exercise is longer than 10 minutes, the maximum amount of energy output is a direct result of the oxidative capacity of the muscle cells. Events lasting 3 to 10 minutes depend on both the anaerobic and the aerobic metabolism. Availability of substrates such as glycogen and triglycerides becomes an important factor in prolonged physical exercise lasting more than 1 to 2 hours : in muscle biopsies taken after a competitive 100 km run, we have found an almost complete depletion of glycogen granules and a 75% decrease in the volume density of intracellular lipid droplets (Oberholzer et al., 1976). Under the same conditions, we have shown a 35-38% decrease in the serum concentrations of most amino acids including alanine, a 44% increase in urea production and no change in 3-methylhistidine production (Décombaz et al., 1979). These findings are compatible with a stimulation of gluconeogenesis, partly at the expense of the amino acid pool, without induction of muscle protein catabolism. Other authors have calculated that the energy supply coming from protein breakdown is of the order of 8-10% under the conditions of long-lasting exercise (Keul, 1972).

The completion of such exceedingly long-lasting efforts as a 100 km run at a sustained pace obviously requires food energy in support to the failing body reserves, primarily to the depleted glycogen stores. The critical importance of a sufficient glucose supply is evidenced by a voluntary food intake during the race directed almost exclusively toward carbohydrate energy (98.4%) (Décombaz et al., 1981a). This may be a more economic way than the breakdown of body protein to provide the necessary glucose and certainly helps reduce the extent of gluconeogenesis.

DIETARY MODIFICATIONS BEFORE EXERCISE

Glycogen stores in skeletal muscle and liver can provide for about 8.5 MJ of energy and hence for a maximum of two hours of heavy physical exercise. Low concentrations of muscle glycogen lead to a marked reduction of performance capacity after 60 minutes of competitive running (Karlsson & Saltin, 1971). Thus, Scandinavian scientists have recommended a special diet for increasing the pre-race muscle glycogen stores (Saltin & Hermansen, 1967). Briefly, the method consists in depleting the glycogen by a heavy training load some days before competition and in eating a carbohydrate-rich diet during the last three days. In an even more sophisticated version, the glycogen depletion is followed by three days of a diet consisting of protein and fat before the planned supercompensation is again induced by three more days with carbohydrate-rich meals. The one or the other type of pre-race carbohydrate loading has been widely accepted in endurance events, e.g. marathon running, cycling, cross-country skiing, etc.

On a long-term basis the average composition of the daily diet has an influence on fuel expenditure during exercise. Experiments have shown that when an adaptation period of a few weeks is allowed after a change to a different diet, the fuels used during exercise tend to be predominantly those provided by the diet (Keys et al., 1950; Phinney et al., 1980). The effects of a change from a mixed diet to either a high-carbohydrate or a low-carbohydrate diet on the exercise respiratory quotient are presented in Table I. The RQ moves in the direction of the "food quotient" and it should be noted that after the low-carbohydrate diet, less muscle glycogen was used during exercise (Phinney et al., 1980). The shift in the RQ indicates a greater or smaller use of carbohydrates during exercise after adaptation to the respective diet, but the nature of the adaptive change is still unknown. Because it is common practice to recommend physical activity together with low-carbohydrate diets in weight reduction programmes, it is of interest to know that exercising on such diets requires less glycogen and burns off lipids more rapidly.

Table I : Long-term diet and fuels for exercise

Dietary carbohydrate supply	Duration	RQ during exercise before diet	after diet
70% (1)	24 weeks	0.87	0.95
5% (2)	6 weeks	0.76	0.66

(1) from Keys et al., 1950; (2) from Phinney et al., 1980

THE PRE-RACE MEAL

Glucose being the main fuel for exercise, it is widely recommended as a pre-race meal. However, the risk of a strong stimulation of insulin secretion followed by a sharp, decrease in blood glucose concentration and even hypoglycaemia has been raised (see, for instance, Ivy et al., 1979). Hypoglycaemia during exercise generally leads to earlier perception of fatigue and impairment of endurance capacity, although some may be less sensitive than others (Felig et al., 1982). Less insulinogenic substrates providing quickly available energy may be suitable alternatives to glucose feedings before excercise. Therefore, in two studies we investigated the effects of pre-exercise meals consisting of either glucose, fructose or medium-chain triglycerides on work capacity, blood concentrations of substrates and metabolites, muscle metabolism and respiratory gas exchange. Both experiments were designed using ^{13}C as a tracer and measuring the excretion of $^{13}CO_2$.

Sixty minutes after ingesting of 1 MJ of either maltodextrins or medium-chain triglycerides, twelve subjects had to exercise on a bicycle ergometer at an intensity eliciting 60% of their maximum aerobic power during one hour. The insulinogenic effect was very marked after carbohydrate administration and it was also followed by a slight hypoglycaemic effect during the exercise period. The medium-chain triglyceride meal had no effect on plasma insulin, and the blood glucose concentrations were more stable throughout the whole time of observation. The calculated contributions of carbohydrate and fat fuels to the energy supply of the working muscles during exercise were the same after the maltodextrin and the medium-chain triglyceride meals. Glycogen depletion in muscle (M. vastus lateralis) averaged 55-58% and was not significantly different after the two test meals. However, the utilization pattern of the exogenously administered substrates was very different. Whereas a considerable part of the medium-chain triglycerides was oxidized already during the rest period between meal ingestion and the start of cycling, oxidation of glucose occurred during the exercise period only. The medium-chain triglycerides caused a moderate ketonemia both at rest and during exercise, but the absorption of one single meal did not influence the contribution of carbohydrates to exercise metabolism. Despite the dietary supply of lipids, the availability of lipid fuels (ketone bodies and fatty acids) to the muscles was not noticeably higher than after the maltodextrin meal. There was no fat-induced sparing effect on the utilization of muscle glycogen, as has been reported after artificially raising the circulating level of free fatty acids (Hickson et al., 1977). Therefore, it seems that an ingestion of medium-chain triglycerides prior to exercise offers no practical advantage (Décombaz et al., 1981b).

In a second experiment, ingestion of a test meal containing 75 g of fructose prevented the glucose-induced increase in plasma insulin concentration and the consequent depression of blood glucose in a similar way to the medium-chain triglycerides. Fructose administration led to a significant increase in blood lactate concentration prior to exercise. Compared with exercise-induced metabolic acidosis, this increase in blood lactate at rest can, however, be neglected, since during exercise the difference between groups disappeared and the performance capacity of our subjects was not at all influenced. Fructose and glucose were oxidized to the same extent, as judged from the amount of $^{13}CO_2$ expired and again, glycogen depletion in M. vastus lateralis was the same after both sugars. Hence fructose seems to be as efficient a fuel for exercise as glucose. In practice, since fructose ingestion before prolonged exercise prevents the occurrence of an insulin-mediated depression of blood glucose, some benefits over glucose ingestion may even be expected in some subjects.

DRUG-LIKE EFFECTS OF NUTRIENTS ON THE ENERGY SUPPLY

Caffeine and nicotinic acid (niacine) are constituents of a normal diet. Strictly, they are no nutrients in the sense that they do not yield energy. When ingested in amounts higher than usual, their pharmacological properties can influence the energy regulation during exercise.

Caffeine

It has been suggested that a glycogen sparing effect could be obtained by influencing the mobilization of free fatty acids, since utilization of the latter substrate in the muscle cell largely depends on its blood concentration. One possible ergogenic aid in this respect would be caffeine. The administration of 330 mg of caffeine one hour before exercise was shown to delay exhaustion by 15 minutes or 19% at an exercise intensity of 80% of the individual maximum aerobic power (Costill et al., 1978). Lipolysis as indicated by blood glycerol concentrations was markedly increased and thus, the longer exercise time was attributed to a reduced rate of glycogen depletion. Beneficial effects of caffeine on endurance have been reported also elsewhere (Ivry et al., 1979; Essig et al., 1980; Berglund & Hemmingson, 1982). However, it should be mentioned that caffeine ingestion has not always been found to have such clear-cut and substantial effects on lipid metabolism and performance. Other authors (Cadarette et al., 1982; Knapik et al., 1982) have demonstrated a stimulation of glycogenolysis that was as important as the influence on lipolysis. In addition, although the stimulation of the nervous system by caffeine is well-documented (Ritchie, 1975), its contribution to endurance capacity through centrally mediated effects is not clearly understood. At present, a possible benefit from caffeine for exercise cannot be attributed solely to its stimulation of fat metabolism.

Nicotinic acid

Lipolysis is strongly inhibited by the administration of nicotinic acid, which then results in an increased carbohydrate utilization during exercise and a reduced performance capacity in endurance events (Bergström et al., 1969). The results of an experiment on rats are summarized in Table II. Trained rats were given 0.6 g/kg nicotinic acid (NIC) or water (CONT) one hour before exercise on a treadmill. Blood indices of lipolysis were reduced in the NIC-group from the onset of exercise, and during the first 40 minutes of work more liver and muscle glycogen was used in the treated animals than in the controls. The reduced delivery of free fatty acids to the working muscles resulted in a faster exhaustion of the glycogen stores. Consequently, the endurance time at a preset pace was shorter after nicotinic acid administration (Décombaz & Roux, 1982).

Table II : Effects of nicotinic acid on lipid metabolism
and endurance in the rat
(all differences statistically significant at the 0.05 level)

	CONT	NIC
Endurance (min.)	140	86
Free fatty acids (μM)	581	445
Glycerol (μM)	132	58
Hydroxybutyrate (μM)	178	58
Glycogen depletion during the first 40 min. :		
Liver (mg/g)	26	40
M. soleus (mg/g)	2.7	3.8

CONCLUDING REMARKS

The energetic regulation of fuel utilization during exercise depends largely on the intensity of the work. If the load is not too heavy, it also depends on the customary diet. However, pre-race meals resulting in quite distinct metabolic patterns at rest usually have little effect on the energy metabolism during exercise.

The success in competition can only partly be influenced by nutritional factors discussed on this article, although some people still believe in miraculous diets. First of all, the athlete needs a well-balanced regimen. For long-distance events, he can under certain circumstances make profit of dietary strategies, such as glycogen loading. Top ·performances in sport, however, are primarily a consequence of genetic endowment, talent and strenuous training over many years.

ACKNOWLEDGEMENTS

The contribution of MM. M.J. Arnaud and G. Philippossian for tracer experiments, and that of Mr. H. Milon for breath acetone measurements is greatly appreciated.

REFERENCES

Berglund, B. & Hemmingsson, P. (1982) Effects of caffeine ingestion on exercise performance at low and high altitudes in cross-country skiers. Int. J. Sports Med. 3:234-236.

Bergström, J., Hultman, E., Jorfeldt,·L., Pernow, B. and Wahren, J. (1969) Effect of nicotinic acid on physical work capacity and on metabolism of muscle glycogen in man. J. Appl. Physiol. 26:170-176.

Cadarette, E.S., Berube, C.L., Evans, W.G., Levine, L. & Postner, B.M. (1982) Effects of varied doses of caffeine on submaximal exercise for exhaustion (poster presentation). Vth Internat. Symposium on the Biochemistry of Exercise, Boston, MA, June 1-5.

Costill, D.L., Dalsky, G.P. & Fink, W.J. (1978) Effects of caffeine ingestion on metabolism and exercise performance. Med. Sci. Sports 10:155-158.

Décombaz, J., Reinhardt, P., Anantharaman, K., von Glutz, G. & Poortmans,
J.R. (1979) Biochemical changes in a 100 km run : free amino acids, urea and
creatinine. Eur. J. Appl. Physiol. 41:61-72.

Décombaz, J., Chiesa, A., von Glutz, G. & Howald, H. (1981a) Nutritional
follow-up of a 100 km footrace. Int. J. Vitamin. Nutr. Res. 51:193-194.

Décombaz, J., Arnaud, M.J., Bur, H., Dalan, E., Milon, H., Moesch, H.,
Philippossian, G., Roux, L., Thélin, A.L. & Howald, H. (1981b) Metabolism of
medium-chain triglycerides during exercise in man. XIIth Internat. Congress
of Nutrition, San Diego, USA.

Décombaz, J. & Roux, L. (1982) Nicotinic acid increases glycogen utilization
and reduces endurance. Int. J. Vitam. Nutr. Res. 52:221.

Essig, D., Costill, D.L. & van Handel, P.J. (1980) Effects of caffeine
ingestion on utilization of muscle glycogen and lipid during leg ergometer
cycling. Int. J. Sports Med. 1:86-90.

Felig, P., Cherif, A., Minagawa, A. & Wahren, J. (1982) Hypoglycemia during
prolonged exercise in normal men. N. Engl. J. Med. 306:895-944.

Hickson, R.C., Rennie, M.J., Conlee, R.K., Winder, W.W. & Holloszy, J.O.
(1977) Effects of increased plasma fatty acids on glycogen utilization and
endurance. J. Appl. Physiol. 43:829-833.

Ivy, J.L., Costill, D.L., Fink W.J. & Lower, R.W. (1979) Influence of
caffeine and carbohydrate feedings on endurance performance. Med. Sci.
Sports 11:6-11.

Karlsson, J. & Saltin, B. (1971) Diet, muscle glycogen, and endurance
performance. J. Appl. Physiol. 31:203-206.

Keul, J. (1972) Energy Metabolism of Human Muscle. In : Medicine and Sport
(Ed. Jokl, E.), Vol. 7, S. Karger, Basel, pp. 52-202.

Keys, A., Brozek, J., Henschel, A., Mickelson, O. & Taylor, H.L. (1950) The
Biology of Human Starvation. Minneapolis, University of Minnesota Press,
pp. 74 and 717.

Knapik, J.J., Jones, B.H., Toner, M.M., Daniels, W.L. & Evans, W.G. (1982)
Influence of caffeine on serum substrates changes during running in trained
and untrained individuals (poster presentation). Vth Internat. Symposium on
the Biochemistry of Exercise, Boston, MA, June 1-5.

Nöcker, J. (1974) Die Ernährung des Sportlers. Hofmann-Verlag, Schorndorf b.
Stuttgart.

Oberholzer, F., Claassen, H., Moesch, H. & Howald, H. (1976) Ultrastruk-
turelle, biochemische und energetische Analyse einer extremen Dauerleistung
(100 km-Lauf). Schweiz. Z. Sportmed. 24:71-98.

Phinney, S., Horton, E.S., Sims, E.A.H., Hansen, J.S., Danforth, E. & La Grange B.M. (1980) Capacity for moderate exercise in obese subjects after adaptation to a hydrocaloric diet. J. Clin. Invest. 66:1152-1161.

Ritchie, J.M. (1975) Ch. XIX. Central Nervous System Stimulants (continued). The Xanthines. In : The Pharmacological Basis of Therapeutics (Ed. Goodman, L.S. & Gilman, A.) Macmillan Co., New York, pp. 367-378.

Saltin, B. & Hermansen, L. (1967) Glycogen stores and prolonged severe exercise. In : Nutrition and Physical Activity (Ed. Blix, B.) Almquist and Wiksell, Stockholm, pp. 32-46.

PROTEIN TURNOVER, NITROGEN BALANCE AND REHABILITATION

Edward B. FERN* and John C. WATERLOW

Clinical Nutrition & Metabolism Unit, Department of Human Nutrition,
London School of Hygiene & Tropical Medicine, London, England

SUMMARY

Not many studies have been done on protein turnover during recovery from malnutrition. Some relevant information can, however, be obtained from measurements on normal growing animals, since rehabilitation and normal growth have in common a rapid rate of net protein synthesis. The key question is the extent to which net gain in protein results from an increase in synthesis or a decrease in breakdown or both.

Different studies have used different methods, and all methods for measuring protein turnover have some disadvantages and sources of error. It is important to bear this in mind in evaluating the results. Consequently, part of this paper will be devoted to questions of methodology.

---

* E.B. Fern is on secondment from the Research Department, Nestlé Products Technical Assistance Co. Ltd., P.O.Box 88, CH-1814 La Tour-de-Peilz, Switzerland.

Whole body protein turnover has been measured in children
recovering from severe malnutrition. During the phase of rapid
catch-up growth the rate of protein synthesis is increased. As
might be expected, it increases linearly with the rate of weight
gain. At the same time there is a smaller increase in the rate
of protein breakdown. The resultant of these two processes is
that, over and above the basal rate of protein synthesis,
1.4 grams of protein have to be synthesized for 1 gram to be
laid down. Very similar results have been obtained in rapidly
growing young pigs.

Experimental studies on muscle growth in general confirm the
conclusion that, at least in muscle, rapid growth is associated
with rapid rates of protein breakdown as well as of synthesis.
This has been shown in muscles of young growing rats, as well as
in muscles in which hypertrophy has been induced by stretch or
other stimuli. In contrast, the evidence suggests that rapid
growth involves a fall in the rate of protein degradation.

The magnitude of the nitrogen balance under any conditions is
determined by the difference between synthesis and breakdown. In
the absence of any storage of amino acids, this must be the same
as the difference between intake and excretion $(S - B = I - E)$.
A question of great interest is whether, at a given intake, the
extent of N balance is determined primarily by regulation of
synthesis and breakdown or by regulation of amino acid oxidation. Clearly, a reduction in amino acid degradation is equivalent to an increase in amino acid intake. An interesting subject
for future research is the extent to which the amino acid
degrading enzymes adapt to the requirements imposed by growth
and rehabilitation.

* * *

In the steady state, body protein content is maintained constant by two
separate, but linked, equilibria (Fig. 1).

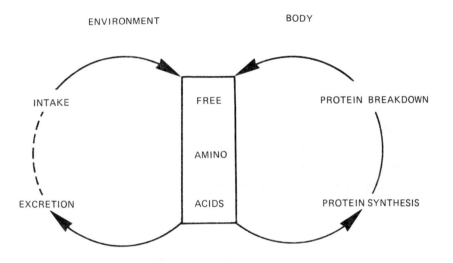

<u>Fig. 1</u> : Equilibria in protein content (steady state)

(i) The familiar nitrogen balance, where nitrogen "in" equals nitrogen "out";

(ii) the balance of protein turnover, where synthesis equals breakdown.

These two balances are separate, in that there may be a greater or lesser input-output flow with no change in turnover, and <u>vice versa</u>. They are linked in that, if one cycle is out of balance, the other must be also. The link of course is through the free amino acid pool. If this pool remains constant, then the amounts of nitrogen leaving and entering it, i.e. the flux, must be equal. Thus :

$$Q = S + E = B + I$$

Where Q is the rate of flux, S and B the rates of synthesis and breakdown and E and I the rates of excretion and dietary intake. The picture looks deceptively simple but the real difficulties begin when we question how these two interlinked cycles are controlled.

The problem can be divided into two parts : first, what actually happens when the steady state is perturbed, as for example by an increased intake of protein or by the physiological demands of growth ? Secondly, once we can describe the events which occur, we have to find out how they are brought about.

The first part of the question, the description of changes in protein turnover under different conditions, has occupied us at the Clinical Nutrition and Metabolism Unit, in London, for many years. It is, therefore, necessary to say something about the methods by which rates of nitrogen metabolism in the whole body are measured.

METHODS

Theoretically, the most valid method at the present time is probably by constant infusion of $1-^{13}C$ leucine and measurement of the specific activity of keto-leucine in plasma, which is taken to represent that of the amino acid at the site of both protein synthesis and oxidation. This method, which has been successfully developed by Matthews and coworkers in St.-Louis (Matthews et al., 1980), is nevertheless technically complex. It requires measurement of the precursor specific activity by gas chromatography-mass spectrometry (GCMS) and the collection and measurement of expired CO_2. We regard this as a research method and have been more concerned in the development of a simple technique which can be more easily applied in different physiological and clinical conditions.

The earliest attempt to measure the rate of whole-body protein turnover was conducted with the stable isotope of nitrogen, ^{15}N (Sprinson & Rittenberg, 1949). Turnover rates were calculated from the time course of the fall in isotope enrichment of urinary urea and were based on the classical concept of first order exchange between compartments. In our view this concept is open to criticism; moreover, the method requires analysis of many samples and small errors can greatly influence the results.

Almost twenty years ago, Picou and Taylor-Roberts (1969), working on children in Jamaica, tried to approach the problem of measurement from a different angle. The idea was to produce an isotopic steady state by continuous infusion of a tracer (^{15}N glycine) and to calculate the turnover rate from the level of labelling in urinary urea when the steady state had been achieved. This is the so-called stochastic or "black box" approach which is independent of any compartmental model.

The practical problem with this method is that because the urea pool in the body is large and turns over slowly ($T\frac{1}{2}$ around 5-10 hours), it takes between 1-2 days to obtain plateau labelling, or even longer on low protein diets when the rate of urea excretion is low.

To make this method more useful for general clinical use, the protocol was modified in two ways (Waterlow et al., 1978b). The first was to use ammonia, rather than urea, as the end-product, primarily because the ammonia pool is small and turns over very fast. The second modification was to use a single dose as opposed to a continuous infusion of tracer. This is not a reversion to the original compartmental method of Sprinson & Rittenberg (1949) because the rate of turnover is not calculated from the slope of the labelling curve but from the area underneath it - a difference of fundamental importance. The theory and calculations of this single dose approach are discussed elsewhere (Waterlow et al., 1978a; 1978b). In practice, all the modified protocol involved was administration of a single dose of labelled amino acid, followed by three 3-hour urine collections and measurement of the ^{15}N enrichment of the ammonia contained within them. The simplicity and the relative speed of this procedure allows for simultaneous measurement of many subjects and an ability to repeat measurements of the rate of protein turnover at intervals of two days or so. On one occasion, as many as fourteen consecutive tests were carried out on a patient given different diets (Garlick et al., 1980a).

Despite the advantages of the single dose method based on urinary ammonia, it was found that the turnover rates measured with this end-product were, in general, lower than those measured with urea (Golden & Waterlow, 1977; Waterlow et al., 1978b). This discrepancy in the rates indicated by the two end-products did not correspond with the basic assumption that there is a single homogeneous pool of metabolic nitrogen in the body which is the precursor for the synthesis of both protein and urinary end-products. Recently this discrepancy has been investigated in our department (Fern et al., 1981). With a further modification of the above single dose method, we have been able to obtain estimates of nitrogen flux from ammonia as well as urea over a 9-hour experimental period. The estimate from urea was made possible by allowing for isotope retained within the urea pool of the body at the end of the period of measurements. This was done on the assumption that the concentration and ^{15}N enrichment of urea in the body was represented by that of plasma water and the volume of the urea pool was equal to that of total body water. The study, which used ^{15}N glycine as the tracer, established that the relationship between the rates of turnover given by ammonia and by urea was an inverse one; when one was high, the other was low. This relationship implied that the assumed pool of metabolic nitrogen in the body was, in effect, divided into two functional pools - one supplying nitrogen for urea synthesis and the other supplying nitrogen for ammonia formation (Fig. 2). The division of the metabolic pool is essentially a physical one reflecting that urea and ammonia are produced in different organs, namely the liver and kidney, and also that absorption of dietary nitrogen occurs at a specific location in the body.

If only one pool of metabolic nitrogen existed, then the total flux of nitrogen in the body would equal the flux of that pool. In a two-pool model, however, the total flux is the sum of the individual fluxes in each of the two compartments. The main difficulty in using this model is that theoretically, it is necessary to know the exact proportion in which the dose of ^{15}N is distributed between the two precursor pools. The problem can be overcome to some extent by calculating a mean value from the rates given by ammonia and urea. We have been using two types of average in this context. The first is a simple arithmetic mean which indicates a rate of nitrogen flux based on an equal distribution of isotope between the two compartments irrespective of their pool size. The second type is a harmonic average, which gives a rate of flux based on a distribution of tracer proportional to the size of each pool. The two averages, in practice, give similar rates of turnover. This can be seen in Table I.

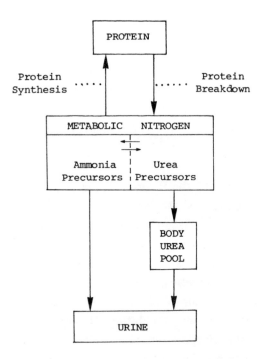

Fig. 2 : Compartmentation of metabolic nitrogen

Table I : Whole-body nitrogen flux measured from the labelling
of ammonia and urea
(g of Nitrogen / 9 h; mean ± SEM)

^{15}N glycine route	End-product			
	Ammonia	Urea	Arithmetic average[a]	Harmonic average[b]
Intravenous	14.7 ± 1.8	30.2 ± 1.4	22.4 ± 0.9	19.5 ± 1.3
Oral	18.6 ± 1.0	24.1 ± 0.7	21.3 ± 0.5	20.7 ± 0.6

(a) Arithmetic average = (QA + QU)/2
(b) Harmonic average   = 2(1/QA + 1/QU)

Table I gives the estimated rate of nitrogen flux derived from both ammonia
and urea and also the two end-product averages, in a study of fed normal
volunteers given ^{15}N glycine either intravenously or orally. The tracer
was administered as a single dose. The results show that the calculated
rates of flux from ammonia and urea varied appreciably with the route of
administration of the tracer. In contrast, those obtained by averaging the
rates given by the two end-products were not significantly different when
the isotope was given intravenously or orally.

One major concern about the use of single amino acid tracers for the
measurement of whole-body nitrogen flux is the degree to which the specific
metabolism of the amino acid represents that of total nitrogen in the body.
This is often referred to as metabolic compartmentation. It is a fundamental
assumption of the stochastic methods that the partitioning of isotope
between the synthesis of protein and excretory products is the same as that
of total body nitrogen. Theoretically, the best available tracer to fulfil
such a requirement would be a protein in which all the nitrogen is labelled.
We have recently measured the rate of nitrogen flux with such a tracer -
uniformly labelled ^{15}N wheat protein - and compared it to rates obtained
with different ^{15}N labelled amino acids. The results are show in Tables II
and III. Table II compares the rates obtained in four male subjects after
giving ^{15}N wheat or ^{15}glycine. As with physical compartmentation, the
rates indicated by ammonia and urea varied considerably. However, the
end-product averages for each tracer were very similar, suggesting that
glycine in studies of this nature is representative of total nitrogen in the
body. This was not so for all amino acids. Table III gives the values for
the harmonic averages obtained in one male subject for different ^{15}N
tracers, including uniformly labelled wheat and labelled glycine. As might
be expected glutamine and alanine, both specific donors of nitrogen to
urinary ammonia and urea synthesis, gave rise to low estimates of flux
relative to that obtained with wheat. On the other hand, lysine and leucine,
whose nitrogen makes only a small direct contribution to ammonia and urea
formation, produced higher estimates of flux - very high in the case of
lysine. Glutamate and aspartate gave intermediate values but, as seen in
Table II, the rate obtained with glycine was almost identical with that from
wheat. This increases our confidence that, for an adult in a fed state,
glycine is a good choice of tracer. In children, especially premature

infants, the situation may be different because of the increased requirement
of amino acids for growth (Jackson & Golden, 1981).

Table II : Whole-body nitrogen flux measured from the the labelling
of ammonia and urea
(g of Nitrogen / 12 h; mean ± SEM)

Tracer	End product			
	Ammonia	Urea	Arithmetic average	Harmonic average
^{15}N glycine	26.8 ± 1.0	30.4 ± 1.8	28.6 ± 1.0	28.3 ± 1.0
^{15}N wheat	35.3 ± 1.6	23.1 ± 1.4	29.2 ± 1.1	27.8 ± 1.2

Table III : Rate of nitrogen flux as indicated by different ^{15}N
labelled tracers
(g of Nitrogen / 12 h)

^{15}N Tracer	Harmonic average
Lysine	113.0
Leucine	37.3
Wheat protein	27.5
Glycine	27.2
Aspartate	25.8
Glutamate	24.7
Glutamine	16.7
Alanine	16.4

This is the current position we have reached in the development of a rapid and practicable method for the routine measurement of protein turnover. There is still some further progress to be made but at least there is now a basis for a working protocol.

PROTEIN TURNOVER AND GROWTH

Growth represents net nitrogen retention, if we think in terms of the conventional nitrogen balance, or, in terms of protein turnover, a higher rate of synthesis than a breakdown. In man, postnatal growth is normally so slow that it is difficult to investigate its mechanism. The normal child at

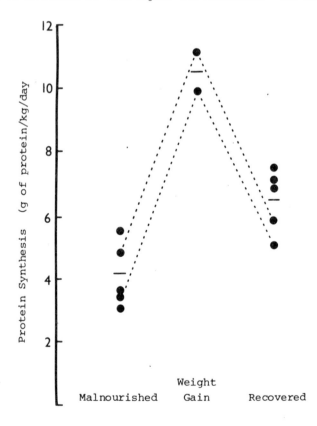

Fig. 3 : Protein synthesis during nutritional rehabilitation of malnourished children.

1 year has a daily weight increment of 0.1-0.2%, with a net nitrogen retention of 25-50 mg/kg/day. It is quite difficult to measure this by the ordinary balance technique, and the difference between protein synthesis and breakdown is too small to be determined with by any confidence by the methods currently available. However, the child recovering from severe malnutrition may be growing at 10-20 times the normal rate and, consequently, provides a better opportunity for studying the mechanism of protein gain. Fig. 3 shows the results obtained in such children in Jamaica (Golden et al., 1977a). During the rapid growth phase of nutritional rehabilitation, the rate of body protein synthesis was doubled.

In theory, net protein synthesis can be produced by either an increase in the rate of synthesis or a decrease in breakdown, or a combination of both. In fact, it seems rather surprisingly, that rapid growth is accompanied by an increase in the rate of breakdown as well as of synthesis. Fig. 4, again

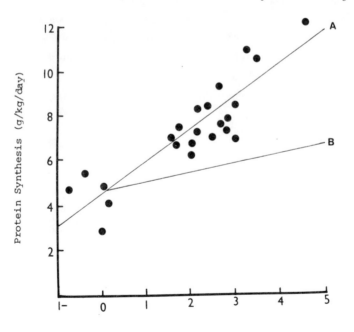

Protein Gain (g/kg/day)

Fig. 4 : Protein synthesis as a function of protein gain during nutritional rehabilitation of malnourished children. Line A : rate of whole body synthesis; Line B : rate of breakdown.

taken from studies in Jamaica on children recovering from malnutrition (Waterlow & Jackson, 1981), shows (Line A) that the more rapid the rate of protein gain, i.e. growth, the greater the rate of whole-body protein synthesis. Line B represents the corresponding rate of breakdown, obtained by substracting the net gain from the total rate of synthesis. This line too has a positive slope, which means that the greater the rate of growth the faster the breakdown rate. The slopes show that for every gram of protein gained, 1.4 g are synthesized, of which 0.4 g is broken down again. Similar figures have been obtained in the young pig (Reeds et al., 1980) which is growing at about the same fractional rate as the recovering child.

Protein synthesis requires substantial amounts of energy. In another series of studies on malnourished children, oxygen consumption was measured after a test meal. The magnitude of the postprandial increase in oxygen uptake was linearly related to the rate of growth (Brooke & Ashworth, 1972). This suggested that, perhaps, the increased oxygen consumption represented the energy cost of growth, stimulated by food. Later work comparing protein turnover in the fed and fasted state has shown that the rate of whole-body protein synthesis responds with surprising sensitivity to the intake or withdrawal of food. This has been shown in studies in which the changes in rate have been measured with both ^{14}C leucine (Clugston & Garlick, 1982a) and ^{15}N glycine (Fern et al., 1981).

Normal growth must represent a balanced gain of protein in all tissues according to a predetermined pattern. Nevertheless, it does not necessarily follow that the mechanism is going to be the same in all tissues. In man, it is obviously extremely difficult to measure rates of protein turnover in individual tissues, and we, therefore, have to resort to studies in animals. Table IV shows some results obtained in young and adult rats (Waterlow et al., 1978a). It is evident that in the young growing rat, the fractional synthesis rate of skeletal muscle protein is treble that of the adults, whereas in liver and kidney the rates are similar at both ages.

Since in the growing animal, net deposition of protein is presumably occur-ring in liver and kidney as well as in muscle, growth in these tissues must have been achieved by a reduction in breakdown rather than an increase in synthesis. There is indeed a good deal of evidence to suggest that in liver the rate of protein breakdown is quite sensitive to nutritional and hormonal stimuli, whereas in muscle it is the rate of synthesis which is most responsive.

Table IV : Rates of protein synthesis in tissues of
the young and adult rat

	Fractional synthesis rate (% per day) Mean ± SD	
	Young	Adult
Liver	57 ± 12	54 ± 6
Kidney	50 ± 10	51 ± 7
Brain	17 ± 3	11 ± 2
Heart	19 ± 4	11 ± 2
Skeletal muscle	15 ± 3	5 ± 1

Other studies (Bates & Millward, 1981) have shown that the very high rate of protein synthesis found in skeletal muscle of the rapidly growing weanling rat is accompanied by a high rate of protein breakdown, just as in the recovering children. Similary Laurent et al. (1978) have reported that when muscles in the wing of a chicken are made to hypertrophy, by loading with weights, there is an increase in protein synthesis as well as an increase in protein breakdown. In fact, it seems to be a general rule that hypertrophy of muscle; however, it is brought about, is accompanied by an anabolic increase in protein breakdown. We can but speculate on the functional reason for this increased destruction and renewal, during growth, of the highly complicated myofibrillar system.

ADAPTATION TO LOW PROTEIN INTAKES

Since this symposium is concerned principally with man, we may consider first the effect of low protein intakes on protein turnover in the whole body. The picture at the present time is not entirely clear, partly because when the early work was done, we did not realize the extreme sensitivity of protein turnover to the immediate food intake, the effect of which has to be dissected out from the prevailing diet. In early studies with ^{15}N on children (Golden et al., 1977b) and adults (Steffee et al., 1976), a comparison was made between the effects of adequate and low protein intakes. On low protein, there was no change in the whole-body synthesis rate but a small increase in the apparent rate of breakdown. In a later study on children (Jackson, unpublished) synthesis was reduced on a diet which was only marginally adequate in protein, compared with the rate found on a more generous intake.

Two recent studies with carbon-labelled leucine allow a comparison of the effects both of the "prevailing" protein intake and of the immediate consumption of protein at the time of the test, when the rate of turnover was being measured (Motil et al., 1981; Clugston & Garlick, 1982b). Although their design is very different in both studies, when measurements were made in the fasting state, rates of both synthesis and breakdown were greater with higher prevailing protein intakes. Thus a high protein diet appears to "speed up" protein turnover. In both studies, feeding protein during the test caused some fall in the breakdown rate compared with the post-absorptive state, at all levels of prevailing protein intake. There was, however, disagreement about the effect of immediate feeding on protein synthesis : in the one case the effects were variable, in the other the synthesis rate increased.

A question of great interest is whether a period of inadequate protein intake can cause an adaptive change in the response to a protein meal. Some information on this is provided by the results of Clugston & Garlick (1982b) represented in Table V. The number of subjects is small, so the interpretation must be tentative, but it looks as if a person adapting to a protein free diet not only has a reduced rate of whole-body protein turnover, but

Table V : Effect of feeding, fasting and refeeding after fasting
on protein synthesis and breakdown in obese women

Diet Energy Prot. (kcal) (g)			Synthesis	Breakdown
			g protein / 12 hours	
500	50	Fed	120	77
0	0	Fasting	92	112
500	0	Fed	72	79
0	0	Fasting	70	77
500	50	Refed	74	37

responds to the restoration of protein in the food with a fall in breakdown rather than an increase in synthesis. This is the most economical way of producing net retention.

As with growth, we have to turn to animal experiments to find out the contributions of individual tissues to these changes. The results obtained up to 1978 on rats malnourished and subsequently refed have been summarized elsewhere (Waterlow et al., 1978a) and for reasons of space will not be considered here. Much of this earlier work needs to be repeated with more up-to-date methods because of potentially serious problems associated with the precursor pool of amino acids for protein synthesis (Fern & Garlick, 1974). A recent method for measuring the rate of protein synthesis in individual tissues involving a large flooding dose of tritiated phenylalanine (Garlick et al., 1980b) may prove useful in overcoming the problems of amino acid compartmentation.

It is likely that some of the disagreements in the literature arise not only from differences in the method used to measure protein turnover but also from differences in experimental plan. The exact timing of the measurements is of great importance. Many years ago, it was shown (Mendes & Waterlow, 1958) that when rats are refed after several weeks on a low protein diet, there is an immediate spurt of growth in the liver but a lag of one to two days in muscle. It may well be that there is a similar dissociation in time between changes in the rates of protein synthesis and breakdown and this possibility needs to be incorporated into the experiment.

CONCLUSION

It is evident that changes in the rate of protein turnover do occur both at the tissue level and that of the whole body, in response to many stimuli including alterations in nutritional and hormonal status, infection, physical trauma and in exercise. They have been documented over the last ten to fifteen years with different methods and different types of experimental approach. Perhaps what is now required, in the light of practical experience and understanding of the methods, is to go back and re-investigate some of these situations. This apparent retrogression may well be a necessary step to establish, first of all, the precise extent to which protein metabolism responds directly to these stimuli and, secondly, to obtain more information on subsidiary factors, such as those relating to hormones and energy metabolism.

Accurate measurement is the first step towards understanding the mechanism by which the components of protein turnover are integrated in each tissue and in the body as a whole. The problem for the future is to relate physiological events in the whole animal to biological regulations at the level of the cell.

REFERENCES

Bates, P.C. & Millward, D.J. (1981) Characteristics of skeletal muscle growth and protein turnover in a fast-growing rat strain. Br. J. Nutr. 46:7-13.

Brooke, O.G. & Ashworth, A. (1972) The influence of malnutrition on the postprandial metabolic rate and respiratory quotient. Br. J. Nutr. 27: 407-415.

Clugston, G.A. & Garlick, P.J. (1982a) The response of protein and energy metabolism to food intake in lean and obese man. Hum. Nutr. Clin. Nutr. 36C:57-70.

Clugston, G.A. & Garlick, P.J. (1982b) The response of whole-body protein turnover to feeding in obese subjects given a protein-free diet for three weeks. Hum. Nutr. Clin. Nutr. 36C:391-397.

Fern, E.B. & Garlick, P.J. (1974) The specific radioactivity of the tissue free amino acid pool as a basis for measuring the rate of protein synthesis in the rat in vivo. Biochem. J. 142:413-419.

Fern, E.B., Garlick, P.J., McNurlan, M.A. & Waterlow, J.C. (1981) The excretion of isotope in urea and ammonia for estimating protein turnover in man with ^{15}N glycine. Clin. Sci. 61:217-228.

Garlick, P.J., Clugston, G.A. & Waterlow, J.C. (1980a) Influence of low energy diets on whole-body protein turnover in obese subjects. Am. J. Physiol. 238:E235-E244.

Garlick, P.J., McNurlan, M.A. & Preedy, V.R. (1980b) A rapid and convenient technique for measuring the rate of protein synthesis in tissues by injection of ^{3}H phenylalanine. Biochem. J. 192:719-723.

Golden, M.H.N. & Waterlow, J.C. (1977) Total protein synthesis in elderly people : A comparison of results with ^{15}N/glycine and ^{14}C/leucine. Clin. Sci. & Mol. Med. 53:277-288.

Golden, M.H.N., Waterlow, J.C. & Picou, D. (1977a) Protein turnover, synthesis and breakdown before and after recovery from protein-energy malnutrition. Clin. Sci. & Mol. Med. 53:473-477.

Golden, M.H.N., Waterlow, J.C. & Picou, D. (1977b) The relationship between dietary intake, weight change, nitrogen balance, and protein turnover in man. Am. J. Clin. Nutr. 30:1345-1348.

Jackson, A.A. & Golden, M.H.N. (1981) Interrelationships of amino acid pools and protein turnover. In : Nutrition Metabolism in Man (Ed. Waterlow, J.C. & Stephen, J.M.L.) Applied Science Publishers, London & New Jersey, pp. 361-373.

Laurent, G.J., Sparrow, M.A. & Millward, D.J. (1978) Turnover of muscle protein in the fowl - Changes in the rate of protein synthesis and breakdown during hypertrophy of the anterior and posterior latissimus dorsi muscles. Biochem. J. 176:407-417.

Matthews, D.E., Motil, K.J., Rohrbaugh, D.K., Burke, J.F., Young, V.R. & Bier, D.M. (1980) Measurements of leucine metabolism in man from a primed continuous infusion of L - 1-^{13}C leucine. Am. J. Physiol. 238:E473-E479.

Mendes, C.B. & Waterlow, J.C. (1958) The effect of a low-protein diet and of refeeding on the composition of the liver and muscle in the weanling rat. Br. J. Nutr. 12:74-88.

Motil, K.J., Matthews, D.E., Bier, D.M., Burke, J.F., Munro, H.N. & Young, V.R. (1981) Whole-body leucine and lysine metabolism : Response to dietary protein intake in young men. Am. J. Physiol. 240:E712-E721.

Picou, D. & Taylor-Roberts, T. (1969) The measurement of total protein synthesis and catabolism and nitrogen turnover in infants in different nutritional states receiving different amounts of dietary protein. Clin. Sci. 36:283-296.

Reeds, P.J., Cadenhead, A., Fuller, M.F., Lobley, G.E. & McDonald, J.D. (1980) Protein turnover in growing pigs. Effects of age and food intake. Br. J. Nutr. 43:445-455.

Sprinson, D.B. & Rittenberg, D. (1949) The rate of interaction of the amino acids of the diet with the tissue proteins. J. Biol. Chem. 180:715-716.

Steffee, W.P., Goldmith, R.S., Pencharz, P.B., Scrimshaw, N.S. & Young, V.R. (1976) Dietary protein intake and dynamic aspects of whole-body protein turnover in obese subjects. Metabolism 25:281-297.

Waterlow, J.C. & Jackson, A.A. (1981) Nutrition and protein turnover in man. Br. Med. Bull. 37:5-10.

Waterlow, J.C., Garlick, P.J. & Millward, D.J. (1978a) Protein Turnover in Mammalian Tissues and in the Whole Body. Elsevier/North Holland, Amsterdam.

Waterlow, J.C., Golden, M.H.N. & Garlick, P.J. (1978b) Protein turnover in man measured with ^{15}N : Comparison of end products and dose regimes. Am. J. Physiol. 235:E165-E174.

AMINO ACID SIGNALS AND FOOD INTAKE AND PREFERENCE : RELATION TO BODY PROTEIN METABOLISM[1,2]

Alfred E. HARPER and John C. PETERS

University of Wisconsin-Madison, Department of Biochemistry, College of Agricultural and Life Sciences, Madison, Wisconsin, U.S.A.

SUMMARY

Depressed food consumption is an early response of experimental animals to :

1) a dietary deficiency of either protein or an individual indispensable amino acid; 2) a distortion of the dietary pattern of amino acids when protein intake is low; and 3) a substantial elevation in the protein content of the diet. In each of these conditions the change in feeding behaviour is associated with alterations in concentrations of amino acids in blood but in none of them has the biochemical basis for the depressed food intake been established.

---

1 The research reported in this paper was supported by the College of Agricultural and Life Sciences of the University of Wisconsin-Madison and by U.S. Public Health Service grants AM 10747 and AM 10748 from the National Institutes of Health, Bethesda, MD.

2 The following abbreviations are used in this paper : 5-HT = 5-hydroxytryptamine (serotonin); 5-HIAA = 5-hydroxyindoleacetic acid; NE = norepinephrine; DA = dopamine; DOPAC = dihydroxyphenylacetic acid; HVA = homovanillic acid; IAA = indispensable amino acids.

The depressed food intake of rats consuming a low protein diet in which an imbalance of amino acids has been created by adding quantities of amino acids other than the one most limiting for growth, is associated with elevations in the plasma concentrations of amino acids added to create the imbalance and usually with a depression in the plasma concentration of the growth-limiting amino acid. These changes, in turn, are associated with depression of the concentration of the growth-limiting amino acid in the brain free amino acid pool. Studies in which uptake of amino acids into brain slices has been examined support the conclusion that various distortions of the plasma amino acid pattern, as the result of dietary imbalances of amino acids, can lead to depletion of the brain pool of a specific amino acid through competition between it and other amino acids in surplus in plasma for uptake into brain. The results of studies with rats in vivo of the effects on brain amino acid pools of ingestion of diets containing supplements of amino acids that compete with the growth-limiting amino acid for uptake into brain also support this conclusion.

Depletion of the brain pool of the limiting amino acid as the result of feeding a diet with an amino acid imbalance can be related to overall body protein metabolism. In the young growing animal, protein synthesis is stimulated after a meal. Thus, when the diet is limiting in a single amino acid, that amino acid will be depleted from the circulating body pool. At the same time, the activities of amino acid degrading enzymes are low in animals fed a low protein diet; hence, such animals have limited capacity to degrade surpluses of amino acids. These conditions, depletion of the blood pool of the limiting amino acid and slow removal of surpluses of competing amino acids from the blood, will increase the extent of competition between other amino acids and the limiting amino acid for uptake into brain. The relationship between depletion of the brain pool of the limiting amino acid and the signal for depressed food intake under these conditions has not been established.

The depressed food intake of rats fed a high protein diet is associated with substantial elevations of blood amino acid concentrations, particularly of the large neutral amino acids. As elevations in the concentrations of large neutral amino acids in blood can suppress uptake of tryptophan into brain, it has been postulated that protein consumption may be responsive to changes in brain tryptophan concentration and that this response may be mediated by changes in brain concentration of serotonin, a neurotransmitter formed from tryptophan. Studies of associations among protein consumption and brain tryptophan and serotonin concentrations both in rats fed single diets differing in protein content and in rats offered a choice between two diets differing in protein content, have not supported this hypothesis. Measurements of brain amino acid concentrations of rats at various times during and after adaptation to a high protein intake suggest that brain total free amino acid concentration may play a role in the control of protein intake and, hence, of food intake.

The degree of food intake depression observed in rats fed diets differing in protein content appears to depend on the capacity of the body to degrade amino acids. When food intake is depressed initially in response to increased dietary protein content, brain total indispensable amino acid concentration is about 2.5 mM. In rats that have been fed previously a diet that is low in protein, and therefore have a limited ability to degrade amino acids, this concentration of brain amino acids is associated with depressed food intake. If rats are allowed to become adapted to a high protein intake, their capacity for amino acid degradation (activities of amino acid-degrading enzymes) increases substantially. After adaptation, they can consume much larger quantities of protein before brain indispensable amino acid concentrations reach 2.5 mM. Also, rats that have been adapted to low protein diets and are then fed a high protein meal tend to reject high protein diets when they are subsequently offered a choice between low and high protein diets. Rats that have been adapted to high protein diets, and have the capacity to clear amino acids rapidly from blood, select diets higher in protein content under these conditions. The correlation between protein consumption and brain amino acid concentrations is such as to suggest that total body amino acid-degrading capacity determines the extent to which brain amino acid concentrations rise in response to changes in dietary protein intake and that the extent of this rise may be a factor in the control of subsequent protein consumption.

* * *

The protein content and the amino acid composition of the diet are two of many nutritional variables that can influence food intake and diet selection of experimental animals. Food intake of the young growing rat is depressed when the protein content of the diet is appreciably less than, or greatly in excess of, the protein requirement; when the diet provides an inadequate amount of an indispensable amino acid; or when the proportions of dietary amino acids, especially in a low protein diet, deviate appreciably from the proportional requirements of the rat for amino acids (Harper, 1967).

Also, when the rat is offered two diets that differ widely in protein content, it will select between them to obtain an adequate to somewhat above adequate intake of protein (Mitchell & Mendel, 1921; Anderson, 1979). It will also, when it is offered two diets that differ appreciably in amino acid composition, show a decided preference for one with a well-balanced dietary pattern of amino acids (Harper et al., 1970). If the rat is offered a protein-free diet together with one containing reasonably well-balanced protein, it will select mainly the diet containing protein (Harper, 1967) and will eat an appreciable amount of the protein-free diet only if the protein content of the alternative diet is considerably in excess of the requirement (Musten et al., 1974; Nemetz, 1979). It will, however, select a protein-free diet in preference to one in which the proportions of amino acids deviate markedly from the proportional requirements for amino acids (Harper et al., 1970).

The many observations on effects of alterations in the protein content and amino acid composition of the diet on feeding behavior have led to the inference that some physiological system for the control of food and protein intake in animals responds to differences in the amounts and patterns of amino acids in the food consumed. A further inference that is justified from these observations is that some component of this system enables animals to distinguish among diets that differ in protein content and amino acid composition and to select between two or more diets to obtain a balanced pattern of amino acids and an appropriate amount of protein.

The low food consumption of rats fed a low protein diet would appear to be determined by systems for the control of both protein and caloric intake (Harper & Boyle, 1976). Growth is limited by a low protein intake and, if animals were to consume food that is low in protein primarily to meet protein needs for growth, they would have to consume an excessive amount of calories which would accumulate as fat. The importance of calorie utilization in control of food intake under these conditions is evident from observations that rats provided with an opportunity to dissipate energy, e.g., through cold exposure, will consume more food and protein and grow at a greater rate than those not provided with such an opportunity. If they have a choice, however, rats that are unable to dissipate extra energy will select for a level of protein intake that will permit rapid growth.

Observations on effects of amino acid deficiencies on food intake and blood amino acid concentrations led Frazier et al. (1947) and Almquist (1954) to propose that alterations in blood amino acid concentrations might give rise to a signal to curtail food consumption. Mellinkoff (1957) came to a similar conclusion from studies of associations between food intake and blood amino acid concentrations in human subjects in different physiological and pathological states. Associations between depressed food intake and alterations of the blood amino acid pattern, were also observed in experiments on rats consuming diets in which the amino acid pattern was severely unbalanced. These observations, too, suggested that distortion of the blood amino acid pattern might, in some way, give rise to a signal that would lead to depression of food intake and preference for a diet with a balanced pattern of amino acids or for a protein-free diet, i.e. selection of diets that would restore the blood amino acid pattern to a balanced state (Harper et al., 1970).

The depressed food intake of rats consuming a high protein diet is associated initially with a greatly elevated concentration of total indispensable amino acids in blood and, particularly, with elevations in the concentrations of the branched-chain amino acids (Anderson et al., 1968; Peng et al., 1972). These observations suggested that high blood concentrations of total amino acids might, in themselves, initiate a signal for curtailment of food intake. They might also initiate a signal for selection of a diet that provided less protein, if the choice were available (Harper & Peters, 1981).

Blood amino acid concentrations do not merely reflect the composition of the diet consumed but are the resultant of the quantity and pattern of amino acids ingested and the action of metabolic processes in the body that remove absorbed amino acids (Harper, 1974a). The rate of utilization of amino acids by an organism is influenced, in turn, by its physiologic and nutritional state (Harper, 1974b). Thus, if the concentrations and patterns of amino acids in blood give rise to signals which are monitored by a system that controls feeding behavior, the feeding response of an animal to a particular dietary regimen would be expected to change if a change in its nutritional or physiologic state affected enzymes or hormones involved in amino acid metabolism.

When animals that have become adjusted to a low protein intake are fed a diet with a high protein content, they undergo metabolic adaptations that increase their capacity to degrade amino acids. As these adaptations occur, the rate at which they are able to clear amino acids from the blood increases and the amount of protein they will consume voluntarily, rises (Anderson et al., 1968). When rats consuming a diet in which there is an imbalance of amino acids, and whose food intake is severely depressed, are given a small supplement of the limiting amino acid, their capacity for protein synthesis is increased, their blood amino acid pattern is restored toward normal, and their food consumption increases (Sanahuja & Harper, 1963). Thus, the extent to which a modification of the protein content or amino acid composition of a diet will elicit a signal leading to an alteration in food intake or preference depends upon the protein metabolic state of the organism and its ability to maintain homeostasis of blood amino acid concentrations. Some of these relationships are discussed in the sections below on effects of "dietary disproportions" and "altered dietary protein content" on feeding behaviour of the rat.

DIETARY DISPROPORTIONS OF AMINO ACIDS AND FOOD INTAKE AND PREFERENCE

This subject has been reviewed in considerable detail (Leung & Rogers, 1969; Harper et al., 1970; Rogers & Leung, 1973; Harper, 1974c; Harper & Boyle, 1976; Rogers, 1976; Rogers & Leung, 1977) so the essence of the basic observations will be outlined only briefly.

Interest in the subject arose from observations that growth of rats consuming a low protein diet to which had been added a mixture of indispensable amino acids devoid of the one that was limiting for growth, was depressed below that of rats fed the unsupplemented low protein diet. This effect was attributed to a dietary imbalance of amino acids. Subsequently rats that were offered the opportunity to choose between a protein-free diet and a diet in which there was an imbalance of amino acids were found to select mainly the protein-free diet on which they could not survive. They would also show a strong preference for a diet with a balanced amino acid pattern. If they were offered only the imbalanced diet, their food intake was depressed, within 2 to 4 hr, below that of rats offered the balanced control diet.

The food intake depression was associated with changes in the pattern of plasma amino acids. The concentration of the limiting amino acid, which was not included in the amino acid mixture added to the control diet to create the imbalance, was depressed; the plasma concentrations of amino acids that were included in the mixture were elevated. This resulted in a large increase in the ratios of the concentrations of several of the amino acids in plasma to that of the limiting amino acid. The time of occurrence of the

depression in food intake corresponded closely with the time at which alterations in the plasma amino acid pattern were observed. This suggested that
the changes in plasma concentrations of amino acids were the source of a
signal that led to the depressed intake and the preference for a protein-
free diet or a diet with a balanced pattern of amino acids, consumption of
either of which would restore the plasma pattern toward the standard state.

These responses to unbalanced dietary amino acid patterns indicate that
some amino acid-sensitive feeding system can evidently override systems that
ordinarily control energy intake. If, however, rats are fed an amino acid-
imbalanced diet in a cold environment which greatly increases energy
expenditure, their food intake is not depressed. They will, nevertheless,
still select a diet with a balanced, over one with an imbalanced, amino acid
pattern under these conditions (Harper & Boyle, 1976). Also, the high food
intake of obese (ob/ob) mice was not depressed when they were fed a
threonine-imbalanced diet, indicating a reduced sensitivity to amino acid
signals (Tews & Harper, 1982). Thus, despite the striking responses observed
in feeding behaviour to altered dietary amino acid patterns, it should be
recognized that under some conditions, other control systems will predomi-
nate over those responding to amino acid signals.

The central nervous system has long been recognized as the site of a
mechanism for control of food intake. Bilateral lesions in the ventromedial
hypothalamus result in hyperphagia, presumably owing to damage to nerves
involved in eliciting a signal for satiety. However, food intake of rats
with such lesions was depressed when they were fed a diet in which there was
an imbalance of amino acids, indicating that this response was not dependent
on an intact hypothalamus. Subsequently, Rogers & Leung (1973) demonstrated
that food intake of rats with bilateral lesions of the prepyriform cortex or
certain parts of the amygdala was not depressed when they consumed a diet
containing disproportionate amounts of amino acids. (These lesions did not
prevent the food intake depression resulting from ingestion of a high protein diet). They also found that infusion of a small amount of the limiting
amino acid directly into the carotid artery would prevent the depression of
food intake in intact rats (Rogers & Leung, 1977). These observations
indicated that nerve tracts in the brain distinct from those in the

hypothalamus, were involved in the feeding responses elicited by diets containing disproportionate amounts of amino acids.

In studies in our laboratory, Peng et al. (1972) observed that when rats were fed a low protein diet to which had been added a surplus of all of the indispensable amino acids except histidine, the concentration of histidine in brain fell more rapidly and to a greater extent than in plasma. This suggested that both the severity of depletion of the brain pool of histidine and the altered feeding behaviour under these conditions were associated with metabolic responses in organs and tissues other than brain.

In earlier studies in which rats fed either the control or the histidine-imbalanced diet had consumed equal quantities of histidine, the plasma amino acid pattern of those consuming the imbalanced diet was more out of balance than had been anticipated (Sanahuja & Harper, 1963). Plasma histidine concentration was depressed below that of the controls despite the equal histidine intakes of the two groups. Also, amino acids that were added to create the imbalance were not cleared rapidly from the blood. Studies done with isotopically-labelled amino acids provided no evidence that depletion of the plasma pool of the limiting amino acid, when rats were fed an imbalanced diet, could be accounted for by increased oxidation of the limiting amino acid (Rogers, 1976). However, more of the labelled amino acid was incorporated into the proteins of liver, and probably of other organs, of rats consuming an imbalanced diet than of those consuming the control diet (Ip & Harper, 1974). Increased incorporation of amino acids into tissue proteins of rats fed an imbalanced diet would contribute to depletion of the plasma pool of the limiting amino acid. Also, in other studies, Kang-Lee & Harper (1977) found that when protein intake, and hence the activities of amino acid-degrading enzymes, are low, as they are in rats fed the diets used in the studies of effects of dietary disproportions of amino acids on feeding behavior, the rate of clearance of a load of an amino acid from blood is slow.

These two metabolic responses, depletion of the plasma pool of the limiting amino acid and a slow rate of removal of other amino acids that were present in elevated concentrations in plasma, would account for the high ratio of the plasma concentrations of amino acids included in the diet to create an imbalance to that of the limiting amino acid. This type of distortion of the plasma amino acid pattern would create the potential for competition between the limiting amino acid and amino acids in surplus in plasma with which it shared a common transport carrier for uptake into brain. Competition of this type could inhibit uptake of the limiting amino acid and, thereby, lead to its depletion from the brain amino acid pool.

Although much information about competition among amino acids for entry into brain was available before these observations were made, most of it had been obtained with extremely high concentrations of the amino acids used as competitors (see Tews et al., 1978 for references). We, therefore, examined the extent to which amino acids in surplus in plasma of rats fed imbalanced diets competed with the limiting amino acid for uptake into brain slices. The large neutral amino acids (phe, tyr, met, leu, ile, val) competed with tryptophan and histidine for uptake into brain slices at concentrations within the physiologic range (Lutz et al., 1975; Tews et al., 1978). These observations suggested that distortion of the plasma amino acid pattern of rats fed an imbalanced diet might result in depletion of the brain pool of the limiting amino acid as the result of competition among plasma amino acids in vivo for uptake into brain. If depletion of the brain pool initiated a signal for depression of food intake or preference for a diet with a balanced amino acid pattern, one would predict that additions of competing amino acids to the diet of an animal would result in concomitant depressions of food intake and of the brain pool of the limiting amino acid.

In a series of experiments designed to test this prediction, we have observed that additions of large neutral amino acids to a low protein diet limiting in either histidine or tryptophan will result in a depression of food intake which is accompanied by depletion of the limiting amino acid from the brain pool. Both of these effects are prevented by a supplement of the limiting amino acid (Harper, 1977a). Similarly, additions of small neutral amino acids, alanine and serine, to a low protein diet limiting in threonine will result in depressions of both food intake and the brain pool of threonine which are prevented by a supplement of threonine (Tews et al., 1980). Also, additions of the basic amino acids, arginine and ornithine, to

diets limiting in lysine, will depress food intake and the brain pool of
lysine, both of which are prevented by a supplement of lysine (Tews et al.,
1981).

If brain threonine concentration is plotted against the ratio of the
plasma concentrations of threonine to the small neutral amino acids (what
might be called the competition ratio), a high and significant correlation
(Fig. 1) is observed (Harper, in press). Similar relationships are observed
between brain tryptophan or histidine concentration and the ratio of the
plasma concentrations of tryptophan or histidine to large neutral amino
acids (Harper, 1977).

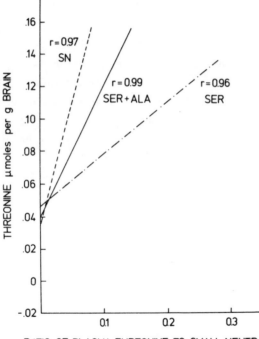

Fig. 1 : Relationship be-
tween brain concentration of
threonine   and   ratio   of
threonine to small neutral
(SN) amino acids in plasma
(after Tews et al., 1980).

RATIO OF PLASMA THREONINE TO SMALL NEUTRAL A A

Taken altogether these observations have led us to conclude that the
protein metabolic state of rats fed low protein diets in which an amino acid
imbalance is created, leads to conditions in which the plasma amino acid
concentrations are altered in such a way as to increase competition for
uptake of the limiting amino acid into brain. As a result, the brain pool of

the limiting amino acid is depleted and this depletion is associated with depression of food intake. The signal that arises from depletion of the brain pool of the limiting amino acid has not been identified but, presumably, it is monitored in the prepyriform cortex (and possibly the amygdala) as lesions in this region prevent food intake depression (Rogers & Leung, 1973). The food intake depression is also prevented by a small supplement of the limiting amino acid. Effects of this type have been observed with rats fed on diets in which histidine, threonine, tryptophan, lysine or isoleucine are limiting. Thus, there does not appear to be unique specificity for a particular amino acid. This also makes it unlikely that the effect is mediated by depletion of a specific brain neurotransmitter as threonine and isoleucine are not known to be precursors of neurotransmitters. Also, a supplement of tryptophan, which is a precursor of serotonin, will prevent food intake depression if the diet is limiting in tryptophan but will contribute to the depression if the diet is limiting in histidine (Harper, 1977a). Whether depletion of the brain pool of the limiting amino acid may affect synthesis of a peptide or protein involved in control of feeding behaviour of animals consuming diets in which there are disproportions of amino acids, remains to be explored.

PROTEIN INTAKE AND SELECTION

In addition to dietary disproportions of amino acids, the protein content of the diet can also influence food consumption and food preference of the rat. In early studies, the ability of the rat to consume an adequate quantity of amino acids when it was offered a choice between proteins differing in nutritive quality suggested the existence of physiological control mechanisms based on a specific appetite for protein (Richter et al., 1938) or preferences for individual proteins (Scott & Quint, 1946). The results of a number of experiments suggest that intake of protein is regulated separately from that of other dietary constituents (Rozin, 1968). Furthermore, it has been proposed that weanling rats allowed to self-select among diets maintain protein intake at a constant proportion of total calories (Musten et al., 1974). The percentage of total calories selected as protein (34% for casein, 43.9% for gluten) was different depending upon the nutritional quality of the protein as reflected by its amino acid composition. In other studies in which rats were offered a choice between two casein diets varying only in methionine or lysine content, the intakes of both methionine and lysine were regulated at levels sufficient to meet the animals' requirements for these amino acids (Muramatsu & Ishida, 1982). From these observations, it is apparent that young rats will select an amount of protein that meets amino acid needs for maximal growth and, in order to obtain a sufficient amount of the limiting amino acid to support growth, will select a greater proportion of total calories from a low than from a high quality protein.

In a series of experiments on weanling rats allowed to self-select between two diets differing widely in protein content, Ashley & Anderson (1975) found that protein intake over a 4 week period was inversely proportional to the ratio of the plasma concentration of tryptophan (Trp) to neutral amino acids (NAA = leu + ile + val + tyr + phe) measured on the final day of the experiment. The relationship differed quantitatively for different proteins presumably because the plasma Trp/NAA ratio is affected by both the amino acid composition of the protein and the amount of protein eaten.

Fernstrom & Wurtman (1972) have shown that brain Trp uptake is influenced by the plasma ratio of Trp/NAA. They (1971) have also shown that the effect of ingestion of carbohydrate on the plasma ratio of Trp/NAA is the opposite of that of protein (carbohydrate ingestion increases the ratio, protein ingestion decreases the ratio); therefore, the effects of these two macronutrients on the uptake of Trp into brain are also opposite. Because brain serotonin (5-HT) synthesis is sensitive to the supply of its precursor Trp, meals containing predominantly protein or carbohydrate also have opposite effects on brain 5-HT concentration. These observations led the authors to suggest that brain 5-HT may play a role in feeding behaviour (Fernstrom & Wurtman, 1971).

The link provided by Fernstrom & Wurtman between the plasma Trp/NAA ratio and brain 5-HT synthesis led Ashley & Anderson (1975) to propose that brain 5-HT may be important specifically in regulating protein intake. They concluded that protein intake should be inversely proportional to brain 5-HT concentration. In applying their hypothesis to the regulation of protein intake, a meal that causes an increase in the plasma Trp/NAA ratio and, hence, in brain 5-HT content, should shift the subsequent preference of the animal to a meal that decreases the plasma Trp/NAA ratio and brain 5-HT concentration.

Studies done in our laboratory (Peters & Harper, 1981) with self-selecting weanling rats, and experiments done by Chee et al. (1981b) on lean and obese mice, have confirmed that an inverse association is observed between protein intake and the plasma ratio of Trp/NAA when measurements extend over a wide range of protein intakes. However, brain 5-HT concentrations in our study (Fig. 2) and in that of Romsos et al. (1982) were not inversely proportional to protein intake over the narrower range of protein intakes of self-selecting rats, as predicted by the Ashley & Anderson hypothesis.

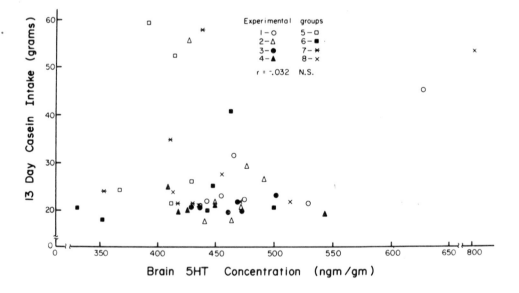

Fig. 2 : Relationship between brain serotonin concentration and protein (casein) intake of individual rats allowed to select between diets containing 15% or 55% casein with or without various amino acid supplements (after Peters & Harper, 1981).

Woodger et al. (1979) have reported that adding supplemental Trp to diets of self-selecting rats elevated brain 5-HT and caused a reduction in protein intake. However, this finding was not consistent as only a very high level of Trp, probably high enough to depress food intake of rats offered a single diet (Harper et al., 1966), added to diets of normal rats reduced protein intake, whereas moderate Trp additions actually increased protein intake.

Experiments in which tryptophan loading was used to elevate brain 5-HT and 5-hydroxyindoleacetic acid (5-HIAA) had either no effect (Weinberger et al., 1978) on total food intake or depressed intake slightly (Gibbons et al., 1981). Our own studies have shown that elevating brain 5-HT and 5-HIAA concentrations by more than 50% had no effect on total food consumption or on protein selection (Peters et al., 1981).

There seems to be little doubt that brain serotoninergic neurons are involved in the feeding response (Blundell, 1977); however, there is contro- versy regarding the specific role of these neurons, and whether factors in the diet have any direct influence on their activity. The dietary studies of Ashley & Anderson (1975) indicate that the correlation between protein intake and brain 5-HT concentration is negative but studies in which the activity of 5-HT neurons was altered with drugs or lesions (Ashley et al., 1979) support a positive correlation. In addition, studies in rats by Wurtman & Wurtman (1979) and in humans by Blundell & Rogers (1980) suggest that brain serotoninergic systems regulate carbohydrate selection and not protein intake.

The results of studies of Peters & Harper (1981) and Romsos et al. (1982) clearly challenge the concept that long-term regulation of protein intake is controlled by brain 5-HT concentrations. In light of these negative findings and considering the paucity of experimental support for the serotonin hypo- thesis, alternative theories must be entertained.

In general, most of the evidence cited in support of a role for brain 5-HT in feeding behaviour has been obtained from experiments in which food intake was modified following physical or pharmacological manipulation of brain serotoninergic neurons. Thus, although the work of Fernstrom & Wurtman

(1972) clearly demonstrates that precursor supply can alter brain 5-HT synthesis, there is, as yet, no firm basis for the contention that diet-induced alteration of brain 5-HT content will modify subsequent feeding behaviour.

It has been observed (Johnson et al., 1979) that regulation of protein intake in self-selecting rats occurs on a meal-to-meal basis. When rats select between two isocaloric diets differing in protein content, any given meal will have a fixed protein to calorie ratio. In an effort to detect possible signals arising from a single meal that might affect protein selection during subsequent meals, we have studied short-term changes in plasma and brain amino acid concentrations and patterns, and brain NE, DA, 5-HT, DOPAC, HVA and 5-HIAA concentrations, in rats that had ingested single meals having a wide range of protein concentrations. Also, as the capacity of the rat to degrade amino acids affects the plasma concentration and pattern of amino acids (Harper, 1977b), we have examined the effects of allowing rats to adapt to levels of dietary protein (casein) ranging from 5% to 75%, on the activities of certain amino acid-catabolizing enzymes in liver, on plasma and brain amino acids, and on brain concentrations of NE, DA, 5-HT and selected metabolites. It was assumed that, taken all together, information from these studies would allow us to predict the effect of a single meal of specific dietary casein content on plasma and brain amino acids and brain neurotransmitters in rats in different protein metabolic states.

In short-term feeding studies rats were adapted to 20% casein diets and trained to consume their entire daily ration in 6 hours during the dark cycle. We then fed different groups of rats (100 g initially) single meals of diets containing 0, 5, 10, 15, 20, 25, 35, 45 or 55% of casein. Blood was collected and brains were taken at 20, 60, 150, 240 and 330 minutes after food was presented.

Compared with the food intake of rats fed moderate levels of protein, that of rats fed protein-free diets or diets containing 35% or more of casein was depressed at all time points. The food intake depressions among rats fed the protein-free diet were associated with levels of both plasma and brain amino acids that were below those of rats fed moderate levels of protein. The decreased food intakes of groups fed diets containing 35% or more of casein were associated with elevations of both plasma and brain concentrations of the 10 indispensable amino acids [IAA] (taken as a sum) above those of rats fed moderate protein levels (about 20%). The maximum concentration of IAAs measured in brain approximated 2.5 mM; this concentration was approached in the groups fed 35%, 45% and 55% casein.

Among the IAAs, the sum of the concentrations of branched-chain amino acids (BCAA) in both plasma and brain closely reflected the amount of protein eaten during the immediately preceding feeding period. This relationship was also seen in our previous self-selection studies (Peters et al., 1981; Peters & Harper, 1981) and in diet selection studies reported by others (Chee et al., 1981b). Brain total dispensable amino acid (DAA = asp, glu, gly, ser, ala, tyr) concentrations (taken as a sum) were not greatly affected by diet and were about 20 mM among all groups.

No consistent relationships were found among brain concentrations of NE, DA, 5-HT, DOPAC, HVA and 5-HIAA, and either total food intake or protein intake. The plasma Trp/NAA ratio was also not linearly related to protein intake; the ratio increased when the casein content was increased up to 10%, was nearly constant when the diet contained from 10% of casein up to and including 35% of casein and declined in rats fed 45% and 55% casein diets. Among the groups fed 10% up to 35% casein diets, the plasma concentration of Trp rose at the same rate with increasing dietary protein content as the sum of NAAs in the denominator, leaving the ratio unchanged.

The results of experiments with rats allowed to adapt to various levels of dietary protein (casein) stress the importance of the amino acid catabolic state of an animal in determining the concentration and pattern of amino acids in both plasma and brain. We fed groups of rats for 11 days, on diets ranging in casein content from 5% to 75% in increments of 5%. Food intake of rats was depressed initially when the diet fed contained 5%, 10% or greater than 35% of casein. For the duration of the experiment, food intake of the groups fed the higher protein diets improved on successive days; the length of depression and the severity of depression depended on the level of protein fed. Rats fed the low levels of protein (5%, 10%) grew poorly and food intake remained depressed. The gradual improvement in growth and food intake of rats fed 35% of casein or greater was accompanied by dramatic increases in the activities of the two liver enzymes measured, serine-threonine dehydratase (SDH) and glutamate-pyruvate transaminase (GPT). Compared with values for rats fed 5% of casein, liver SDH and GPT activities among rats fed 75% casein diets were 180-fold and 8-fold greater, respectively. Liver enzyme activities among rats fed the higher levels of protein (greater than 30%) were elevated in direct proportion to the degree of food intake depression observed during the first days of the experiment.

The enhanced ability of the liver of adapted rats to degrade amino acids was associated with reduced plasma concentrations of those amino acids for which degrading enzymes were induced. Brain amino acid concentrations were similarly reduced. With the exception of the BCAAs, the catabolic enzymes for which are not induced by a high protein intake, brain IAA concentrations of the adapted rats were actually inversely proportional to the protein content of the diet (or protein intake). As a consequence of liver enzyme induction, adapted animals could consume large enough quantities of the high protein diets to satisfy energy demands without experiencing increases in brain IAA concentrations comparable to those that, in unadapted rats, are associated with depressed food consumption.

Brain neurotransmitter or metabolite concentrations did not correlate with food or protein intake on a consistent basis in these long-term studies; but, as in our earlier studies, plasma and brain BCAA concentrations correlated highly with protein intake. The close association between BCAA concentrations in plasma and brain and protein intake, under a wide variety of conditions, suggests that BCAAs could serve, irrespective of the animal's metabolic state, as an indicator of total protein ingestion. Thus, if an animal, previously adapted to a high protein intake, were given the opportunity to select between high and low protein diets, elevated concentrations of plasma and brain BCAAs could provide a signal for the animal to shift from the high to the low protein diet and, thereby, obtain a moderate dietary level of protein. Among rats that were adapted to a high protein diet (50%) and had elevated BCAA concentrations before they were allowed to self-select for protein, we have observed a shift in preference to achieve a diet with a moderate protein content (20%).

As BCAAs compete actively with other large neutral amino acids for uptake into brain, the accumulation of BCAAs in blood plasma and brain extra-cellular fluid when animals ingest a high protein diet, may protect the brain and certain neurons, specifically, from large influxes of amino acids that are precursors of neuroactive compounds. BCAAs have been used clini-cally to prevent an accumulation in brain of neurotransmitter precursor amino acids in patients with hepatic encephalopathy. These patients exhibit depressed plasma concentrations of BCAAs while circulating levels of other large neutral amino acids are elevated, a situation which leads to enhanced uptake of the neurotransmitter precursors, tyrosine, phenylalanine and tryptophan into brain and contributes to derangement of brain function. The coma associated with this type of liver disease can be relieved by paren-teral infusions of BCAA solutions (Bernardini & Fischer, 1982).

Recently we have performed experiments in which we observed that meta-bolic state (amino acid catabolic state) has an influence on protein selection. Young rats were allowed to adapt to diets containing 20% or 50% of casein, or to self-selection between 5% and 55% casein diets (these rats chose 18% protein during adaptation). Each of these groups was separated into two sub-groups, one of which was fed a single small meal containing 5% casein and the other 50% casein. All six groups were then allowed, immedia-tely after the meal, to choose betweeen 5% and 55% casein diets for the subsequent 5 hours.

Rats that were fed a small meal of the 5% casein diet, subsequently selected 33% casein if they had been adapted to the 20% casein diet; 44% casein if they had been adapted to the 50% casein diet; and 32% casein if they had been adapted to self-selection and had been consuming 18% of casein (Table I). Rats that were fed a small meal of the 50% casein diet, subse-quently selected 24% casein if they had been adapted to the 20% casein diet; 20% casein if they had been adapted to the 50% casein diet; and only 13% casein if they had been adapted to self-selection and had been consuming 18% of casein.

Table I : Effect of modifying plasma and brain amino acids patterns
on protein selection of growing rats trained to eat
their daily ration in 6 hours

Small meal pretreatment (4 g in 1 hr)	Adapted to 20% casein		Adapted to 50% casein		Adapted to selection	
	5 hr after small meal	Total*	5 hr after small meal	Total	5 hr after small meal	Total
Percent protein selected 5% vs. 55%						
5% casein	33	21	44	26	32	22
50% casein	24	34	20	29	13	25

* Indicates the total percentage of the diet consumed as casein including
the pretreatment meal.

These results clearly demonstrate that rats in different amino acid
catabolic states are responsive to differences in the protein content of the
diet and are capable of adjusting their protein selection on a meal-to-meal
basis. Thus, rats fed a low protein pre-meal selected protein subsequently
in proportion to their ability to degrade amino acids; the most dramatic
response being the very high percentage of dietary protein selected by rats
that had been adapted to a high protein intake. Our results are similar to
those of Leung et al. (1981) who noted an effect of adaptive metabolic state
on protein selection. An interesting finding was that rats, adapted to
self-selection of protein prior to receiving the 50% casein pre-meal,
selected much less protein subsequently than rats fed the same pre-meal but
previously adapted to single diets. This observation suggests that protein
selection has a learned component such that the animals can associate some
sensory characteristic of each diet with the physiologic response that was
experienced previously.

In every self-selection study done in our laboratory with growing rats
allowed to choose between two protein-containing diets, the level of casein
selected has been about 20% of the total diet. The amino acid content of

this amount of casein approximates the needs of the rat for growth. These observations provide compelling evidence that protein selection by the growing rat is determined in large measure by nutritional needs. If this is the case, then, as amino acid requirements diminish with age, so should the proportion of protein selected.

We tested this assumption in weanling rats allowed to select between 5% and 55% casein diets for 60 days (final body weights exceeded 300 g). Until growth began to slow at about day 40 of the study, the rats selected close to 20% of the diet as casein. After growth began to slow the choices became somewhat erratic, some rats choosing substantially less than 20% (7-10%) others substantially more (40-50%). This behavior persisted throughout the remainder of the study and the average amount of protein selected among the older rats was either comparable to that chosen by younger rats or increased slightly with age.

The fact that in our experiments and in those of Ross et al. (1976), protein selection did not diminish uniformly with age, suggests that nutritional needs are not the sole determinant of protein intake. Instead, it appears that in both young and old self-selecting rats, some minimum protein intake, sufficient to satisfy maintenance and growth requirements, is insured by whatever mechanism controls intake, and that consumption of protein above this level is tolerated only if the animal can catabolize the excess of amino acids. In young rats the variability within a population in the amount of protein selected is usually small, probably because rats electing to eat only a moderate amount above the requirement (10-15%) would need to undergo metabolic adaptations in order to be able to metabolize rapidly the excess of ingested amino acids. In older rats there is a greater difference between the level of protein needed to meet the requirement and the level of protein which would be needed to induce enzymatic adaptations. This is because both the protein and energy requirements are low at maturity and the liver is large, giving the older rat surplus capacity to degrade amino acids. In these animals, individual taste preferences for protein may be the primary determinants of protein intake above the requirement. This could account for the fairly large variation in protein selection within a population of adult rats.

Our information on changes in brain amino acid concentrations with changing protein intake and adaptive state, suggests that rats consume an amount of protein that results in brain IAA concentrations falling within certain limits, what might be called, in analogy to radioisotope measurements, a "window". One attractive feature of the amino acid concentration "window" hypothesis is that it can account for the feeding responses observed both when rats are fed single diets containing different levels of protein and when animals are allowed to self-select for protein. With rats fed single diets, when the concentration of protein in the diet is low, feeding will cause the concentrations of brain IAAs to fall below the tolerated minimum and food intake is curtailed. Food intake depression is also observed when the protein content of the ingested diet is so high that brain amino acid concentrations exceed the upper limit of the "window". If, however, a choice is provided between high and low protein diets, the animal appears to balance its consumption of the diets offered in such a way that brain amino acid concentrations are maintained within the acceptable limits. The level of dietary protein selected by an animal, therefore, need not be precise as long as the resulting brain amino acid concentrations fall within the acceptable range.

A number of factors that have not been mentioned may also influence an animal's selection of dietary protein content. These include palatability (Rolls, 1981), mineral content of the diet (Leprohon et al., 1979), the non-protein energy source (Chee et al., 1981a) and the protein content of the maternal diet (Leprohon & Anderson, 1980). These factors may be responsible in part for the variability of protein selection of rats within a group and between laboratories. However, the fundamental determinant of selection in our studies would seem to be the ability of any given diet to maintain brain amino acid concentrations within an acceptable range. The influence of maternal dietary protein content on the selection of protein by the off-spring suggests that the limits of the acceptable "window" of brain amino acid concentrations may be influenced by the protein feeding behaviour of the mother during pregnancy.

CONCLUSIONS

The preceding discussion has outlined feeding and metabolic responses of rats fed a wide range of diets varying in amino acid content and balance. When rats are fed diets having either a balanced or imbalanced pattern of amino acids, the feeding response is both quantitatively and temporally related to changes in plasma and brain amino acid concentrations and patterns. Also, food intake of rats fed diets with an imbalanced or dis-proportionate pattern of amino acids, or diets differing in protein content, depends on the adaptive metabolic state of the animal. In the unadapted animal, food intake is curtailed when the supply of incoming amino acids exceeds the animal's capacity to degrade those amino acids not utilized for protein synthesis. In the adapted animal with a high capacity to degrade amino acids, despite increased protein intake, plasma and brain amino acid concentrations are maintained within the normal range.

The specific signal influencing the feeding behaviour of rats consuming amino acid imbalanced diets appears to be related to the brain concentration of the growth-limiting amino acid (Rogers & Leung, 1973). Our studies indicate that regulation of the intake of balanced protein by rats fed either single diets or allowed to self-select, is not related to diet-induced changes in brain neurotransmitter concentrations but is related rather to the total concentration of IAAs in brain. More specifically, protein intake appears to be maintained at a level that results in brain IAA concentrations that fall within a tolerable range or "window". The lower limit of the "window" appears to relate closely to the brain IAA concentrations associated with a dietary protein intake that just meets the animal's amino acid requirements. The upper limit of the window is probably the maximum concentration of brain IAAs tolerated by the animal without the need for metabolic adaptations to prevent adverse effects. A specific mechanism that converts information about brain total free IAA content into neuro-chemical signals that modify behaviour cannot be suggested at this time. Rogers & Leung (1977) have reported that elevated brain ammonia content, as the result of infusions of ammonia via the carotid artery, caused food intake depression in rats fed low protein diets. These results suggest that brain ammonia concentration could serve as a secondary signal relating brain IAA content and protein feeding behaviour.

One common feature of the many studies of effects of dietary protein and amino acid content on food intake and preference, is that the observed changes in feeding behaviour are associated with alterations of both plasma and brain amino acid patterns. The nature of the feeding responses and the plasma and brain amino acid patterns differ, however, depending upon the status of body protein metabolism.

REFERENCES

Almquist, H.J. (1954) Utilization of amino acids by chicks. Arch. Biochem. Biophys. 52:197-202.

Anderson, G.H. (1979) Control of protein and energy intake : role of plasma amino acids and brain neurotransmitters. Can. J. Physiol. Pharmacol. 57: 1043-1057.

Anderson, H.L., Benevenga, N.J. & Harper, A.E. (1968) Associations among food and protein intake, serine dehydratase and plasma amino acids. Am. J. Physiol. 214:1008-1013.

Ashley, D.V.M. & Anderson, G.H. (1975) Correlation between the plasma tryptophan to neutral amino acid ratio and protein intake in the self-selecting weanling rat. J. Nutr. 105:1412-1421.

Ashley, D.V.M., Coscina, D.V. & Anderson, G.H. (1979) Selective decrease in protein intake following brain serotonin depletion. Life Sci. 24:973-984.

Bernardini, P. & Fischer, J.E. (1982) Amino acid imbalance and hepatic encephalography. Ann. Rev. Nutr. 2:419-454.

Blundell, J.E. (1977) Is there a role for serotonin (5-hydroxytryptamine) in feeding ? Internat. J. Obesity 1:15-42.

Blundell, J.E. & Rogers, P.J. (1980) Effects of anorexic drugs on food intake, food selection and preferences and hunger motivation and subjective experiences. Appetite 1:151-165.

Chee, K.M., Romsos, D.R. & Bergen, W.G. (1981a) Effect of dietary fat on protein intake regulation in young obese and lean mice. J. Nutr. 111:668-677.

Chee, K.M., Romsos, D.R., Bergen, W.G. & Leveille, G.A. (1981b) Protein intake regulation and nitrogen retention in young obese and lean mice. J. Nutr. 111:58-67.

Fernstrom, J.D. & Wurtman, R.J. (1971) Brain serotonin content : Increase following ingestion of a carbohydrate diet. Science 174:1023-1025.

Fernstrom, J.D. & Wurtman, R.J. (1972) Brain serotonin content : Physiological regulation by plasma neutral amino acids. Science 178:414-416.

Frazier, L.E., Wissler, R.W., Stefler, Ch.H., Woolridge, F.L. & Cannon, P.R.
(1947) Studies in amino acid utilization. I. The dietary utilization of
mixtures of purified amino acids in protein-depleted albino rats. J. Nutr.
33:65-83.

Gibbons, J.L., Barr, G.A., Bridger, W.H. & Leibowitz, S.F. (1981) L-tryptophan's effects on mouse killing, feeding, drinking, locomotion, and brain
serotonin. Pharmac. Biochem. Behav. 15:201-206.

Harper, A.E. (1967) Effects of dietary protein content and amino acid
pattern on food intake and preference. In : Handbook of Physiology,
Section 6. Alimentary Canal Vol. 1. Control of Food and Water Intake.
Washington, Am. Physiol. Soc., pp. 399-410.

Harper, A.E. (1974a) Amino acid requirements and plasma amino acid. In :
Protein Nutrition (Ed. Brown, H.) Springfield Il. C.C. Thomas, pp. 130-179.

Harper, A.E. (1974b) Control mechanisms in amino acid metabolism. In : The
Control of Metabolism (Ed. Sink, J.O.) University Park, PA : The
Pennsylvania State University Press, pp. 49-71.

Harper, A.E. (1974c) Effects of disproportionate amounts of amino acids.
In : Improvement of Protein Nutriture (Eds Harper, A.E. & Hegsted, D.M.)
Washington D.C., Nat. Acad. Sci., pp. 23-63.

Harper, A.E. (1977a) Influence of dietary and plasma amino acid patterns on
brain amino acid concentrations. In : Proc. Conf. on Commonalities in
Substance Abuse and Habitual Behavior. Washington D.C., Nat. Acad. Sci.,
pp. 39-59.

Harper, A.E. (1977b) Animal models : Plasma amino acids and body protein
status. In : Clinical Nutrition Update-Amino Acids (Ed. Greene, H.L.,
Holliday, M.A., Munro, H.N.) Chicago, Il., Amer. Med. Assoc., pp. 111-116.

Harper, A.E. (in press) Dispensable and indispensable amino acid interrelationships. Amino Acids : Metabolism and Medical Applications. (Eds
Blackburn, G.L. Young, V. & Grant, J.) Littleton, MA, Wright-PSG Publishing.

Harper, A.E. & Boyle, P.C. (1976) Nutrients and food intake. In : Appetite
and Food Intake (Ed. Silverstone, T.) Berlin, Dahlem Konferenzen,
pp. 177-206.

Harper, A.E. & Peters, J.C. (1981) Amino acid signals and their integration
with muscle metabolism. In : The Body Weight Regulatory System : Normal and
Disturbed Mechanisms (Ed. Cioffi, L.A., James, W.P.T., Van Itallie, T.B.)
Raven Press, New York, pp. 33-38.

Harper, A.E., Benevenga, N.J. & Wohlhueter, R.M. (1970) Effects of disproportionate amounts of amino acids. Physiol. Rev. 50:428-558.

Harper, A.E, Becker, R.V. & Stucki, W.P. (1966) Some effects of excessive
intakes of indispensable amino acids. Proc. Exp. Biol. Med. 121:695-699.

Ip, C.C.Y. & Harper, A.E. (1974) Liver polysome profiles and protein synthesis in rats fed a threonine-imbalanced diet. J. Nutr. 104:252-263.

Johnson, D.J., Li, E.T.S., Coscina, D.V. & Anderson, G.H. (1979) Different diurnal rhythms of protein and non-protein energy intake by rats. Physiol. Behav. 22:777-780.

Kang-Lee, Y.A. & Harper, A.E. (1977) Effect of histidine intake and hepatic histidase activity on the metabolism of histidine in vivo. J. Nutr. 107: 1427-1443.

Leprohon, C.E. & Anderson, G.H. (1980) Maternal diet affects feeding behavior of self-selecting weanling rats. Physiol. Behav. 24:553-559.

Leprohon, C.E. & Woodger, T.L., Ashley, D.V.M. & Anderson, G.H. (1979) Effect of mineral mixture in diet on protein intake regulation in the weanling rat. J. Nutr. 109:827-831.

Leung, P.M.B. & Rogers, Q.R. (1969) Food intake : regulation by plasma amino acid pattern. Life Sci. 8:1-9.

Leung, P.M.B., Gamble, M.A. & Rogers, Q.R. (1981) Effect of prior protein ingestion on dietary choice of protein and energy in the rat. Nutr. Rep. Int. 24:257-266.

Lutz, J., Tews, J.K. & Harper, A.E. (1975) Simulated amino acid imbalance and histidine transport in rat brain slices. Am. J. Physiol. 229:229-234.

Mellinkoff, S. (1957) Digestive System. Ann. Rev. Physiol. 19:193-196.

Mitchell, H.S. & Mendel, L.M. (1921) Studies in nutrition. The choice between adequate and inadequate diet, as made by rats and mice. Am. J. Physiol. 58:211-225.

Muramatsu, K. & Ishida, M. (1982) Regulation of amino acid intake in the rat : Self-selection of methionine and lysine. J. Nutr. Sci. Vitaminol. 28:149-162.

Musten, B., Peace, D. & Anderson, G.H. (1974) Food intake regulation in the weanling rat : self-selection of protein and energy. J. Nutr. 104:563-572.

Nemetz, D.J. (1979) Studies on control of protein intake. M.S. Thesis, University of Wisconsin-Madison.

Peng, Y., Tews, J.K. & Harper, A.E. (1972) Amino acid imbalance, protein intake and changes in rat brain and plasma amino acids. Am. J. Physiol. 222:314-321.

Peters, J.C. & Harper, A.E. (1981) Protein and energy consumption, plasma amino acid ratios, and brain neurotransmitter concentrations. Physiol. Behav. 27:287-298.

Peters, J.C., Tews, J.K., & Harper, A.E. (1981) L-tryptophan injection fails to alter nutrient selection by rats. Trans. Am. Soc. Neurochem. 12:404 (Abstract).

Richter, C.P., Holt, L.E. & Barelare, B. (1938) Nutritional requirements for normal growth and reproduction in rats studied by the self-selection method. Am. J. Physiol. 122:734-744.

Rogers, Q.R. (1976) The nutritional and metabolic effects of amino acid imbalances. In : Protein Metabolism and Nutrition (Ed. Cole, D.J.A. et al.) London, Butterworth, pp. 279-301.

Rogers, Q.R. & Leung, P.M.B. (1973) The influence of amino acids on the neuroregulation of food intake. Fed. Proc. 32:1709-1719.

Rogers, Q.R. & Leung, P.M.B. (1977) The control of food intake : When and how are amino acids involved ? In : The Chemical Senses and Nutrition (Ed. Kare, M.R., Maller, O.) New York, Academic Press, pp. 213-248.

Rolls, B.J. (1981) Palatability and food preference. In : The Body Weight Regulatory System : Normal and Disturbed Mechanisms (Ed. Cioffi, L.A., James, W.P.T., Van Itallie, T.B.) New York, Raven Press, pp. 271-278.

Romsos, D.R., Chee, K.M. & Bergen, W.G. (1982) Protein intake regulation in adult obese (ob/ob) and lean mice : Effects of non-protein energy source and of supplemental tryptophan. J. Nutr. 112:505-513.

Ross, M.H., Lustbader, E. & Bras, G. (1976) Dietary practices and growth responses as predictors of longevity. Nature 262:548-553.

Rozin, P. (1968) Are carbohydrate and protein intakes separately regulated ? J. Comp. Physiol. Psych. 65:23-29.

Sanahuja, J.C. & Harper, A.E. (1963) Amino acid balance and imbalance. X. Effect of dietary amino acid pattern on plasma amino acid pattern and food intake. Am. J. Physiol. 204:686-690.

Scott, E.M. & Quint, E. (1946) Self-selection of diet. IV. Appetite for protein. J. Nutr. 32:293-301.

Tews, J.K. & Harper, A.E. (1982) Food intake, growth and tissue amino acid concentrations in lean and obese (ob/ob) mice fed a threonine-imbalanced diet. J. Nutr. 112:1673-1681.

Tews, J.K., Bradford, A.M. & Harper, A.E. (1981) Induction of lysine imbalance in rats : Relationships between tissue amino acids and diet. J. Nutr. 111:968-978.

Tews, J.K., Good, S.S. & Harper, A.E. (1978) Transport of threonine and tryptophan by rat brain slices : Relation to other amino acids at concentrations found in plasma. J. Neurochem. 31:581-589.

Tews, J.K., Lee Kim, Y.W. & Harper, A.E. (1980) Induction of threonine imbalance by dispensable amino acids : Relationships between tissue amino acids and diet in rats. J. Nutr. 110:394-408.

Weinberger, S.B., Knapp, S. & Mandell, A.J. (1978) Failure of tryptophan load-induced increases in brain serotonin to alter food intake in the rat. Life Sci. 22:1595-1602.

Woodger, T.L., Sirek, A. & Anderson, G.H. (1979) Diabetes, dietary tryptophan and protein intake regulation in weanling rats. Am. J. Physiol. 236:R307-R311.

Wurtman, J.J. & Wurtman, R.J. (1979) Drugs that enhance central serotoninergic transmission diminish elective carbohydrate consumption by rats. Life Sci. 24:895-904.

FOOD PROCESSING AND STORAGE AS A DETERMINANT OF PROTEIN AND AMINO ACID AVAILABILITY

Richard F. HURRELL and Paul-André FINOT

Nestlé Products Technical Assistance Co. Ltd, Research Department, CH-1814 La Tour-de-Peilz, Switzerland

SUMMARY

Protein is perhaps the most reactive of the major food components. During food processing, the essential amino acids, lysine, tryptophan, methionine and cyst(e)ine, may react with other food components causing a loss in amino acid bioavailability and sometimes a reduction in the digestibility of the whole protein molecule. This review first discusses some recent developments concerning protein-polyphenol reactions, racemization and lysinoalanine formation, and then describes the reactions of proteins and reducing sugars (the Maillard reaction) in greater detail. We report on the chemistry of the Maillard reaction, the nutritional and physiological properties of the newly-formed products and their metabolic transit in the rat.

In practice, the Maillard reaction is by far the most important reaction of food proteins. It is especially important in milk products since these are the only naturally-occurring protein foods with a high content of reducing sugar. Lysine is the most sensitive amino acid to damage during processing and storage and its losses may be of nutritional significance to certain population groups, such as babies, who are often dependent on a single manufactured product as their sole source of nourishment.

* * *

The food processing industry developed because of man's need to preserve basic foodstuffs between harvest by preventing bacterial growth and enzymatic deterioration. The modern food industry still maintains its primary function of food preservation but has also added to the variety and the convenience of its products which are readily available at all times of the year. In addition, new technologies have provided many new fabricated foods such as breakfast cereals, texturized soya and infant formula with the proximate composition of human milk. While it is true that most food manufacturers are more interested in bacteriological safety and organoleptic quality, there is a growing awareness of nutrition in general, and in particular of the possible negative aspects of processing such as loss of nutritive value or toxicity. In this paper, we discuss the influence of food processing and storage on the nutritive value of food proteins, although it is clear that in most instances the positive aspects of processing greatly outweigh the negative (Fig. 1).

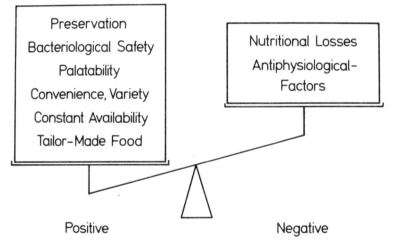

Fig. 1 : The balance between positive and negative aspects of food processing.

During food manufacture, food proteins may be subjected to many different types of process. They may be subjected to physical treatments such as fractionation or heat; biological treatments such as enzymic hydrolysis or fermentation; or chemical treatments such as with alkali, oxidizing agents, sulphur dioxide or solvents. The major processes, however, are those which involve heat treatments to sterilize, cook and dry.

Proteins are the most reactive of the major food components and, apart from the destruction of certain vitamins, reactions between food proteins and other food components are the major chemical reactions that occur during processing. Protein-bound amino acids can react with reducing sugars, poly-phenols, fats and their oxidation products and with various chemical addi-tives such as alkali which has been reported to promote racemization of amino acids and the formation of lysinoalanine. These reactions lead in part to browning and flavour formation (Hurrell, 1982), but can also reduce the nutritional value of the protein (Hurrell, 1980), and in some cases, as with lysinoalanine and advanced Maillard reaction products, they have been reported to cause toxity in rats (Mauron, 1977; 1981).

Lysine, tryptophan, methionine and cysteine are the most reactive amino acids in the protein chain. They are nutritionally essential, or semi-essential in the case of cyst(e)ine, and even when unprocessed, they are often the most limiting amino acids in mixed diets. During processing, they react with other food components either to give covalent complexes or, in the case of methionine, cyst(e)ine and tryptophan, they can also be oxidized (Fig. 2). In order to evaluate the influence of these changes on the nutri-tional value of the protein, three different approaches have been used. These are chemical studies, nutritional (animal) assays and metabolic transit studies.

The chemical studies lead to an understanding of the chemical reactions involved and can give valuable information on the influence of different reaction parameters such as time, temperature, pH, water activity, etc. Analytical tests can determine the chemical loss of each amino acid and sometimes the formation of the modified amino acids. Nutritional studies can indicate the biological availability of the individual amino acids in the newly-formed complexes and demonstrate the influence of processing on protein quality in general and protein digestibility. Metabolic transit studies in rats, using radioactive isotopes, indicate the pathway these newly-formed compounds take in the organism. They can, for instance, give valuable information on whether the amino acid complex is excreted directly in the faeces, whether it is absorbed and afterwards excreted in the urine or on how it is metabolised (Finot, 1982).

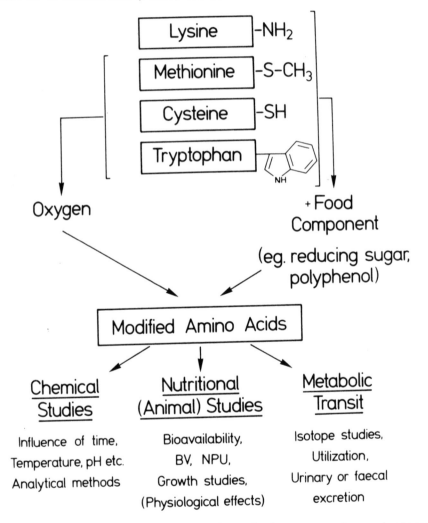

Fig. 2 : Possible amino acid reactions during processing and storage. Experimental design.

For obvious reasons, many of these studies are made using model systems. Often the extreme experimental conditions chosen have little direct relevance to the food manufacturer. However, they do serve to show where the potential problems could occur and subsequent studies with foodstuffs themselves can then help put the problem into perspective. This review first discusses some recent studies and then presents the Maillard reaction in greater detail as this is the major cause of nutritional damage to foodstuffs.

PROTEIN-POLYPHENOL REACTIONS

Phenolic acids such as chlorogenic acid are widely distributed in plants. Sunflower seeds for instance which contain about 16% protein contain up to 2% chlorogenic acid and smaller quantities of caffeic acid and other phenolic acids. During protein extraction, browning reactions take place and coloured products result. In the presence of alkali or polyphenol oxidase, phenolic acids are transformed into the corresponding quinones. These can polymerize to give brown pigments but they can also react with the essential amino acids lysine, methionine and tryptophan (Fig. 3).

Our chemical studies with a model system containing casein and caffeic acid showed that the reaction with lysine depends on pH, oxygen, time and temperature of reaction and concentration of phenolic acid (Hurrell et al., 1982). For our nutritional studies (Fig. 4), an aqueous solution of casein (5%) was reacted with caffeic acid (0.5%) for 3 h at 20°C and the freeze-dried casein-caffeic acid complexes fed to rats. We found losses of available lysine, methionine and tryptophan of 44%, 26% and 13% respectively at pH 10 and 19%, 13% and 8% at pH 7 with tyrosinase (Hurrell et al., 1981). The loss of available tryptophan was probably a result of reduced protein digestibility.

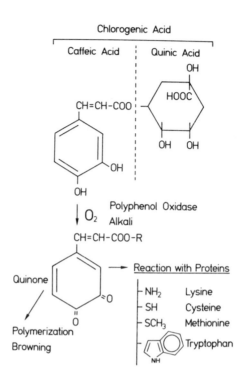

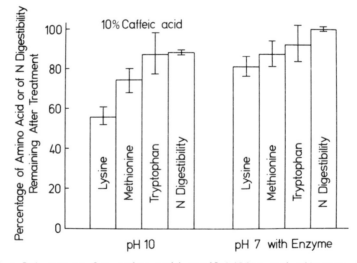

Fig. 3 : Proposed reactions of phenolic acids.

Fig. 4 : Rat assays for amino acid availability and nitrogen digestibility in casein-caffeic acid complexes.

The loss of methionine was probably mainly due to its oxidation to methionine sulphoxide which, under certain conditions, was almost complete. Methionine sulphoxide is slightly less well-utilized by the rat than methionine itself (Cuq et al., 1978). Lysine, however, appeared to form covalent lysine-caffeoquinone complexes and our metabolic study (Fig. 5), using goat's casein labelled with tritiated lysine, indicated that these complexes were not absorbed by the rat but excreted directly in the faeces (Hurrell et al., 1982).

Rats fed on a casein-caffeic acid mixture, containing 4% caffeic acid prepared at pH 10 excreted 26% of the ingested radioactivity in the faeces compared with 12% excreted by rats fed the sample prepared at pH 7 with tyrosinase and 5% by those rats fed the casein control. There were no differences between the percentages of ingested radioactivity excreted in the urine by the rats fed the different diets. From this study, we conclude that proteins extracted from plant material rich in phenolics, especially under alkaline conditions, may be reduced in quality due to protein-poly-phenol browning reactions.

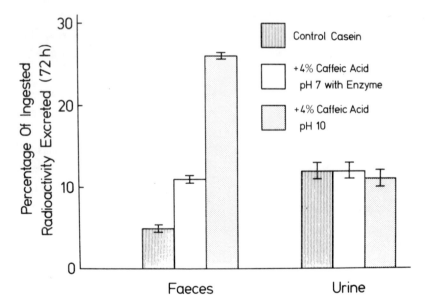

Fig. 5 : Urine and fecal excretions of the radioactivity from rats fed complexes of caffeic acid with casein labelled with tritiated lysine.

RACEMIZATION

There have been several reports recently of the racemization of amino acids during food processing particularly by alkaline treatments (Provansal et al., 1975; Masters & Friedman, 1979) but also after protein-sugar and protein-fat reactions (Hayase et al., 1979). Racemization is thought to proceed by abstraction of the α-proton from the amino acids residue to give a negatively charged carbanion. A proton can then be added back to either side of this optically inactive intermediate thus regenerating the L-form or producing the D-enantiomer.

The absorption of D-amino acids is also much slower than for the corresponding L-form (Gibson & Wiseman, 1951) and, even if digested and absorbed, most D-isomers of essential amino acids are not utilized by man. In his review on the utilization of D-amino acids, Berg (1959) reported that, in man, D-lysine, D-threonine, D-tryptophan, D-leucine, D-isoleucine and D-valine were not utilized. D-phenylalanine could partially replace L-phenylalanine and, of the essential amino acids, only D-methionine was as effective as L-methionine in maintaining N-balance. Kies et al. (1975), Zezulka & Calloway (1976) and Printen et al. (1979), however, have subsequently reported that D-methionine is also poorly utilized by man.

We have recently studied the racemization of amino acids in differently processed proteins (Table I) (Liardon & Hurrell, 1983). In freeze-dried chicken muscle heated 2 h at 121°C, only aspartic acid was racemized to any great extent. Similarly in roasted milk powder aspartic acid was the only amino acid considerably racemized. This however would have a negligible effect on the nutritional value of this product when compared to the almost complete destruction of lysine which occurred under these extreme conditions. There was no significant racemization in milk powders stored and heated at lower temperatures.

Alkaline conditions, on the contrary, were very effective at promoting amino acid racemization and all amino acids were considerably if not completely racemized in a casein solution stirred for 1 h at 80°C in 1M NaOH.

Table I : Percentage of D-amino acids in food proteins
after different treatments

Amino Acid	Chicken muscle 2h at 121°C	Milk powder 20 min. at 230°C	Casein solution 1h at 80°C 0.2M NaOH	1M NaOH
Ala	0	3	20	42
Val	0	0	4	18
Leu	0	2	9	34
Ile	1	0	4	30
Cys	0	0	-	-
Met	0	0	30	43
Phe	0	1	31	44
Lys	0	0	13	41
Asp	15	31	37	51
Glu	0	9	28	42
Ser	1	0	44	54
Thr	0	0	28	34
Tyr	1	0	14	49
His	-	-	36	49

Percentage of D-amino acid = $\frac{D}{D + L}$ x 100

50% D-amino acid = 100% RACEMIZATION.

Alkaline treatments are often used during the processing of plant proteins such as soya to promote better solubility and functional properties. Recently Friedman et al. (1981) reported that racemization increased with time and temperature of treatment particularly at pH values greater than 10. They analysed 5 commercially-prepared products containing soy protein or sodium caseinate, including a soy-based infant formula and reported that 9-17% of the aspartic acid was in the D-form. These values are relatively low and are difficult to evaluate since the amount of other D-amino acids were not reported. Aspartic acid is the most sensitive amino acid to racemization and, although a non-essential amino acid, its racemization could reduce protein digestibility (Hayashi & Kameda, 1980). As well as provoking amino acid racemization, alkaline treatments can destroy certain essential amino acids and further reduce protein digestibility due to the formation of cross-linkages, such as lysinoalanine and lanthionine, in the protein chain. When used in food processing, alkaline treatments should be rigorously controlled.

LYSINOALANINE

Lysinoalanine can be formed in proteins by reaction of lysine with dehydroalanine, produced by cyst(e)ine (Bohak, 1964) and serine degradation (Manson & Carolan, 1972). It was originally detected in alkaline-treated proteins but has since been shown to be present in other heated foodstuffs including milk products (Fritsch & Klostermeyer, 1981).

It is absent from spray-dried powders but, in liquid infant formulas, it can vary from 100-1000 mg per kg crude protein depending in part of the sterilisation process used (Table II). The amount of lysinoalanine formed depends on the severity of the heat treatment. UHT treatment followed by aseptic fill gives much less lysinoalanine than HTST or conventional in-can sterilization.

Lysinoalanine formation in food proteins has been reviewed extensively by Finot (1983). Although its formation has little influence on lysine or cyst(e)ine bioavailability, it is a cross-linkage and would be expected to reduce overall protein digestibility. More importantly however it has been shown to produce renal cytomegaly in rats. This lesion is characterized by enlarged nuclei and increased amounts of cytoplasm in epithelial cells of the straight portion of the proximal tubule. Cytomegaly however has never been found in any other species including mice, rabbits, dogs and monkeys and even in rats it is reversible.

Table II : Lysinoalanine content (mg/kg crude protein)
of milk-based liquid infant formulas

Heat process	Lysinoalanine	
	Mean	Range
UHT/Aseptic fill (3 sec., 150°C)	300	(160-370)
HTST (2.5 min., 124°C)	540	(260-1030)
Conventional sterilization (10 min., 115°C)	710	(410-1030)

A recent development is the demonstration that the dose-response is not the same for the different stereoisomers of lysinoalanine (Feron et al., 1978). The LD isomer is about 10 times more active at provoking cytomegaly than the LL form. If lysinoalanine is formed via the β-elimination reaction and dehydroalanine, then the alanine moiety may be present as either the L or the D isomer since the H atom may add back to either side of the α-carbon atom (Fig. 6). This would give equal parts of LL and LD lysinoalanine. The DL and DD isomers may also be formed in strong alkali if lysine racemizes.

Another proposed mechanism for lysinoalanine formation is a substitution reaction whereby lysine reacts directly with serine phosphate without passing through dehydroalanine (Friedman, 1977). This would generate the less active LL isomer only and is possibly the pathway for lysinoalanine formation in milk. Unfortunately, analytical methods do not exist to distinguish between the different isomers.

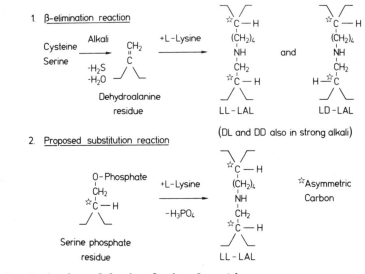

Fig. 6 : Mechanism of lysinoalanine formation.

Lysinoalanine is a toxicological problem, not a nutritional one. Recently, Codex Alimentarius (document CX/VP 82/5) concluded, on the basis of all available evidence on lysinoalanine toxicity, that there was no need to fix limits for the presence of lysinoalanine in vegetable foods. As infant formulas however are the sole source of nourishment for young babies, it would still seem wise that all liquid infant formula receive the least severe sterilization process possible.

MAILLARD REACTION

The reaction of proteins with reducing sugars is the major source of nutritional damage to food proteins during processing and storage. The Maillard reaction occurs during the baking of bread, the production of breakfast cereals, the heating of meats, especially when in contact with vegetables, but most importantly during the processing of milk products since milk is the major naturally-occurring protein food that has a high content of reducing sugar.

"Early" Maillard reaction   "Advanced" Maillard reaction

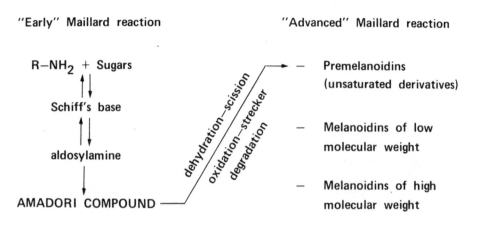

Fig. 7 : Schematic representation of the Maillard reaction.

Maillard reactions appear to follow common pathways (Fig. 7) which can be divided conveniently into early Maillard reactions and advanced Maillard reactions (Hurrell & Carpenter, 1974; Hurrell, 1980). The first step involves the condensation reaction between the carbonyl group of the reducing sugar and the amino group of lysine. The reaction rapidly proceeds via Schiff's bases to the deoxyketosyl compound or Amadori product. This is the major form of blocked lysine after early Maillard reactions. As this stage there is no colour formation.

The brown pigments develop on degradation of the deoxyketosyl compounds and during the advanced Maillard reactions which follow. These reactions are not well-defined but are responsible for the numerous flavours and odours as well as browning, and for a further reduction in protein nutritive value.

The lysine molecule is destroyed during advanced Maillard reactions, as are other essential amino acids such as tryptophan and methionine, possibly by reaction with the active intermediate compounds such as dicarbonyls and aldehydes (Finot et al., 1982; Hurrell et al., 1983). There is a reduction in the bioavailability of all essential amino acids, even leucine which has an unreactive paraffinic side chain. Such findings are explained by the formation of profuse enzyme resistant cross-linkages between various amino

acid side chains via the reactive breakdown products. These cross-linkages
reduce the rate of protein digestion by preventing enzyme penetration or by
masking the sites of enzyme attack.

The results from metabolic transit studies (Finot & Magnenat, 1981) on
the reaction products of lysine and glucose are summarized in Table III.

Table III : Urinary and faecal excretion of early and advanced
            Maillard products (Finot & Magnenat, 1981)

	% ingested radioactivity in :	
	Urine	Faeces
Free fructosyl-lysine	64	14
Protein-bound fructosyl-lysine	11	6
Premelanoidins	27	64
Melanoidins	4	87

Free fructosyl-lysine, the Amadori product, is well-absorbed by the rat.
It is not utilized however and an average of 64% of the ingested dose was
excreted unchanged in the urine. Protein-bound fructosyl-lysine on the other
hand was not well-digested and only 11% was released from the protein,
absorbed and excreted unchanged in the urine. Those units not absorbed are
destroyed by the intestinal flora of the hind-gut, thus explaining the low
recovery of only 6% of the ingested fructosyl-lysine in the faeces.

The advanced Maillard melanoidins and premelanoidins were isolated from a
heated mixture of casein and ^{14}C-glucose by gel filtration and enzymic
hydrolysis. The fraction termed melanoidin was strongly coloured, the pre-
melanoidin was less coloured and of lower molecular weight. The melanoidin
fraction was almost completely indigestible and about 90% was found in the

faeces. The lower molecular weight premelanoidins were partly absorbed and 27% of their radioactivity was detected in the urine. It has been suggested on the basis of rat studies that the premelanoidins may in some way be toxic or growth depressing (Adrian, 1974). The toxicological aspects of Maillard reaction products are further discussed by Mauron (1981) and Lee et al. (1981).

The determination of lysine after Maillard reactions may pose problems and different analytical methods may give widely contrasting results (Hurrell & Carpenter, 1981). Using the furosine technique, however, it has been shown that Maillard reactions in milk products generally do not go beyond the Amadori products (Finot et al., 1981). Table IV shows the percentage of lysine residues blocked as lactulosyl-lysine during the processing of milk.

Table IV : Blockage of lysine as Amadori product in milk products

Heat Process	% Blocked Lysine
Freeze-drying	0
Pasteurization	0
UHT sterilization	0 - 2
Spray-drying	0 - 2
Spray-drying infant formula	5 - 10
HTST sterilization	5 - 10
Conventional sterilization	10 - 15
Roller-drying	20 - 50
Spray-drying lactose-hydrolysed milks or casein-glucose mixtures	15 - 70

There is little or no blocked lysine in freeze-dried, pasteurized, or UHT treated milk or in spray-dried milk powder. Spray-drying of infant formulas however, which contain more lactose, may give up to 10% lysine blockage as does HTST sterilization of liquid milk. Conventional in-can sterilization of liquid milk gives 10-15% blockage and roller-drying 20-50% blockage.

Specialised formulas, such as lactose-hydrolysed milks or formula containing glucose for lactose-intolerant infants, give the greatest problems during drying as glucose is far more reactive than lactose. Conventional spray-drying can give up to 70% blockage in these special formulas, although by careful control of the heat treatment, this can be reduced to around 15%.

Further Maillard reactions can take place during storage of milk powders. These can be minimized by low moisture content of powders, hermetic packaging and cool storage conditions and, in temperate climates, pose no real problems. In hot countries, however, where distribution networks are less well-developed, milk powders may be subjected to high temperatures at some time during their transport and storage. Whole milk powders containing 2.5% moisture were stored for several weeks at 60°C and 70°C and the fate of lysine, methionine, tryptophan (Hurrell et al., 1983; Fig. 8) and most of the B vitamins (Ford et al., 1983; Fig. 9) was investigated.

Storage at 60°C (Fig. 8) is an example of early Maillard damage. After 9 weeks storage, the product still retained its natural colour even though about 40% of the lysine was blocked as lactulosyl-lysine. Methionine and tryptophan were stable.

At 70°C, 50% of the lysine units were blocked as lactulosyl-lysine after only 2 weeks storage in a product which still retained its natural colour. From 3 weeks onwards the product became a deep red brown colour as lactulosyl-lysine degraded and advanced Maillard reactions took place. Available methionine and tryptophan as measured by Streptococcus zymogenes were progressively reduced probably due to an impairement of protein digestibility, although there is some evidence that they may be destroyed by reaction with advanced Maillard reaction products (Finot et al., 1982).

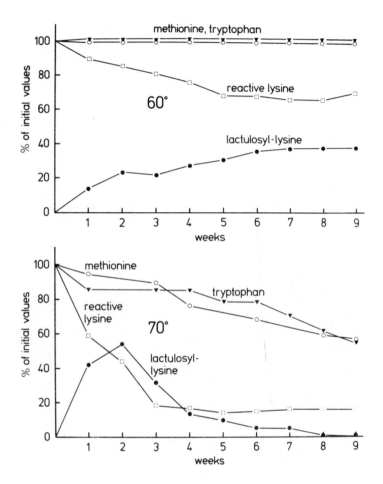

<u>Fig. 8</u> : Amino acid content of milk powders stored at elevated temperatures (Hurrell et al., 1983).

Rather more impressive were the rapid destruction of certain B vitamins at 70°C in parallel with the degradation of lactulosyl-lysine and the appearance of advanced Maillard reaction products (Fig. 9). It appeared that between the second and fourth week of storage, thiamin, vitamin B_6, vitamin B_{12} and pantothenic acid were extensively destroyed due to their reaction with advanced Maillard reaction products. At 60°C, the B vitamins

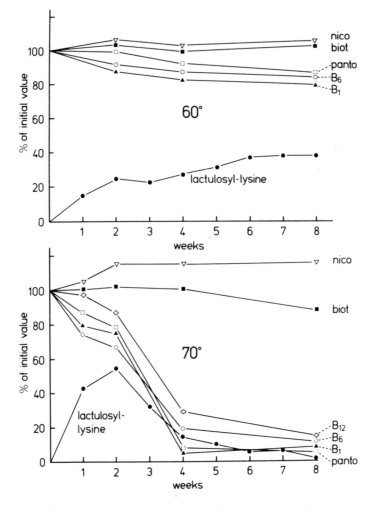

Fig. 9 : B vitamin content of dried milk stored at elevated temperature (Ford et al., 1983).

were relatively stable. Vitamin A, vitamin C and folic acid can also be lost during milk powder storage but their destruction is more closely related to oxygen content rather than the extent of the Maillard reaction. These high vitamin losses at 70°C are somewhat academic since the browned products are clearly organoleptically unacceptable. More important perhaps is the relatively high stability of B vitamins at 60°C and the high loss of lysine at 60°C without any significant change in product colour.

CONCLUSIONS

The importance of these nutritional losses must be evaluated in the diet as a whole. They would appear to have little influence in adult nutrition as most adults obtain their amino acids and proteins from a wide variety of foodstuffs. Indeed, many nutrient losses during heat treatments are unavoidable and would seem to be easily outweighed by the beneficial effects of food processing on food safety, shelf-life and convenience. Nutritional damage, however, may be extremely important if the damaged protein is part of a baby food or infant formula, as infants often rely on a single manufactured product as their sole source of nourishment. Food manufacturers should take special care over the nutrient composition and protein quality of baby foods especially if alkali treatments are involved or if Maillard reactions are possible. If they do not, the negative effects of processing may outweigh the beneficial ones.

REFERENCES

Adrian, J. (1974) Nutritional and physiological consequences of the Maillard reaction. In : World Review of Nutrition and Dietetics, vol. 19, Basel, S. Karger, pp. 71-127.

Berg, C.P. (1959) Utilization of D-amino acids. In : Protein and Amino Acid Nutrition (Ed. Albanese, A.A.) New York, Academic Press, pp. 57-96.

Bohak, Z. (1964) N^{ϵ}-(DL-2-amino-2-carboxyethyl)-L-lysine, a new amino acid formed on alkaline treatment of proteins. J. Biol. Chem. 239:2878-2887.

Cuq, J.L., Besançon, P., Chartier, L. & Cheftel, J.C. (1978) Oxidation of methionine residues of food proteins and nutritional availability of protein-bound methionine sulphoxide. Food Chem. 3:85-102.

Feron, V.J., van Beek, V.J., Slump, L. & Beems, R.B. (1978) Toxicological aspects of alkali treatment of food protein. In : Biochemical Aspects of Protein Food (Ed. Adler-Nissen, J.) 11th FEBS Meeting, vol. 44, pp. 139-147.

Finot, P.A. (1982) Nutritional and metabolic aspects of protein modification during food processing. In : Modification of Proteins Food, Nutritional and Pharmacological Aspects (Ed. Feeney, R.E. & Whitaker, J.R.) Advances in Chemistry Series 198, Washington D.C., Am. Chem. Soc., pp. 91-124.

Finot, P.A. (1983) Lysinoalanine in food proteins. Nutr. Abstr. Rev. Clin. Nutr. 53:67-80.

Finot, P.A. & Magnenat, E. (1981) Metabolic transit of early and advanced Maillard products. In : Progress in Food and Nutrition Science. Maillard Reactions in Food (Ed. Eriksson, C.), vol. 5, Oxford, Pergamon Press, pp. 193-207.

Finot, P.A., Deutsch, R. & Bujard, E. (1981) The extent of the Maillard reaction during the processing of milk. In : Progress in Food and Nutrition Science. Maillard Reactions in Food (Ed. Eriksson, C.), vol. 5, Oxford, Pergamon Press, pp. 345-355.

Finot, P.A., Magnenat, E., Guignard, G. & Hurrell, R.F. (1982) The behaviour of tryptophan during early and advanced Maillard reactions. Int. J. Vitam. Nutr. Res. 52:226.

Ford, J.E., Hurrell, R.F. & Finot, P.A. (1983) Storage of milk powders under adverse conditions. 2. Influence on the content of water-soluble vitamins. Br. J. Nutr. (in press).

Friedman, M. (1977) Crosslinking amino acids - stereo-chemistry and nomen-clature. In : Protein Crosslinking, Nutritional and Medical Consequences. Advances in Experimental Medecine and Biology (Ed. Friedman, M.), vol. 86B, New York, Plenum Press, pp. 1-27.

Friedman, M., Zahnley, J.C. & Masters, P.M. (1981) Relationship between in vitro digestibility of casein and its content of lysinoalanine and D-amino acids. J. Food Sci. 46:127-134.

Fritsch, R.J. & Klostermeyer, H. (1981) Bestandsaufnahme zum Vorkommen von Lysinoalanin in milcheiweisshaltigen Lebensmitteln. Z. Lebensm. Unters. Forsch. 172:440-445.

Gibson, Q.H. & Wiseman, G. (1951) Selective absorption of stereoisomers of amino acids from loops of the small intestine of rat. Biochem. J. 48:426-429.

Hayase, F., Kato, H. & Fujimaki, M. (1979) Racemization of amino acid residues in proteins during roasting. Agr. Biol. Chem. 37:191-192.

Hayashi, R. & Kameda, I. (1980) Racemization of amino acid residues during alkali treatment of proteins and its adverse effect on pepsin digestibility. Agr. Biol. Chem. 44:891-895.

Hurrell, R.F. (1980) Interaction of food components during processing. In : Food and Health : Science and Technology, (Ed. Birch, G.G. & Parker, K.J.) London, Applied Science Publishers, pp. 369-388.

Hurrell, R.F. (1982) Maillard reaction in flavour. In : Food Flavours. Part A : Introduction (Ed. Morton, I.D. & MacLeod, A.J.) Amsterdam, Elsevier Scientific Publishing Company, pp. 399-437.

Hurrell, R.F. & Carpenter, K.J. (1974) Mechanisms of heat damage in proteins. 4. The reactive lysine content of heat-damaged material as measured by different ways. Br. J. Nutr. 32:589-604.

Hurrell, R.F. & Carpenter, K.J. (1981) The estimation of available lysine in foodstuffs after Maillard reactions. In : Progress in Food and Nutrition Science. Maillard Reactions in Food and Nutrition (Ed. Eriksson, C.), vol. 5, Oxford, Pergamon Press, pp. 159-176.

Hurrell, R.F., Finot, P.A., Jaussan, V. & Cuq, J.L. (1981) Nutritional consequences of protein-polyphenol browning reactions. XIIth International Congress of Nutrition, August 16-21, San Diego, California (USA). Abstract No 260.

Hurrell, R.F., Finot, P.A. & Cuq, J.L. (1982) Protein-polyphenol reactions. 1. Nutritional and metabolic consequences of the reaction between oxidized caffeic acid and lysine residues of casein. Br. J. Nutr. 47:191-211.

Hurrell, R.F., Finot, P.A. & Ford, J.E. (1983) Storage of milk powders under adverse conditions. 1. Losses of lysine and of other essential amino acids. Br. J. Nutr. (in press).

Kies, C., Fox, A. & Aprahamian, S. (1975) Comparative value of L-, DL-, and D-methionine supplementation of an oat-based diet for humans. J. Nutr. 105: 809-814.

Lee, T.C., Kimiagar, M., Pintauro, S.J. & Chichester, C.O. (1981) Physiological and safety aspects of Maillard browning of foods. In : Progress in Food and Nutrition Science. Maillard Reactions in Food (Ed. Eriksson, C.), vol. 5, Oxford, Pergamon Press, pp. 243-256.

Liardon, R. & Hurrell, R.F. (1983) Amino acid racemization in heated and alkali-treated proteins. J. Agric. Food Chem. 31:432-437.

Manson, W. & Carolan, T. (1972) The alkali-induced elimination of phosphate from β-casein. J. Dairy Res. 39:189-194.

Masters, P.M. & Friedman, M. (1979) Racemization of amino acids in alkali-treated food proteins. Agr. Food Chem. 27:507-511.

Mauron, J. (1977) General principles involved in measuring specific damage of food components during thermal processes. In : Physical, Chemical and Biological Changes in Food caused by Thermal Processing (Ed. Hoyem, T. & Kvale, O.) London, Applied Science Publishers, pp. 328-359.

Mauron, J. (1981) The Maillard reaction in food : a critical review from the nutritional standpoint. In : Progress in Food and Nutrition Science. Maillard Reactions in Food (Ed. Eriksson, C.), vol. 5, Oxford, Pergamon Press, pp. 5-35.

Printen, K.J., Brummel, M.C., Cho, E.S. & Stegink, L.D. (1979) Utilization of D-methionine during total parenteral nutrition in postsurgical patients. Am. J. Clin. Nutr. 32:1200-1205.

Provansal, M.P., Cuq, J.L. & Cheftel, J.C. (1975) Chemical and nutritional modifications of sunflower proteins due to alkaline processing. Formation of amino acid crosslinks and isomerization of lysine residues. Agr. Food Chem. 23:938-943.

Zezulka, A.Y. & Calloway, D.H. (1976) Nitrogen retention in men fed isolated soybean protein supplemented with L-methionine, D-methionine, N-acetyl-L-methionine or inorganic sulfate. J. Nutr. 106:1286-1291.

ENERGY/PROTEIN INTERRELATION IN EXPERIMENTAL FOOD RESTRICTION

Krishna ANANTHARAMAN

Nestlé Products Technical Assistance Co. Ltd, Research Department,
CH-1814 La Tour-de-Peilz, Switzerland

SUMMARY

The effects of restriction of energy and/or protein intake were studied in rats during pregnancy and lactation and in adult mice. Three approaches were employed : sucrose stimulus-induced reduction in the intake of an adequate diet given simultaneously; restriction of intake of a complete food by 30% of ad libitum levels and selective protein or energy restriction of a high protein diet and a non-protein diet. Casein or lactalbumin was the protein source. During lactation the rat's natural food intake regulatory mechanism prevailed over the sucrose stimulus. Restricted intake of the complete food and selective restriction of protein or energy, variably influenced gestational and lactational performance and weight of young at weaning. A distinct regulation in the intake of protein and energy during pregnancy and lactation was observed on selective energy or protein restriction. Restricted feeding of the composite diet starting in the second year of life, to mice previously ad libitum fed a lab chow of constant composition, promoted the highest survival rate.

* * *

Chronic overfeeding starting early in life leads to precocious and excessive growth including early skeletal and sexual maturity as well as early integration of central nervous system (Kennedy & Pearce, 1958; Widdowson & Kennedy, 1962). Furthermore, it has been observed that such early overnutrition may lead to obesity and related metabolic disorders which negatively influence life-span (Berg & Simms, 1960; Ross, 1959; Ross et al., 1976; Silberberg & Silberberg, 1955; Stuchlikova et al., 1975).

In fact to date the only proven way of actually extending life-span is by restricting food intake (McCay et al., 1939, 1943; Barrows & Roeder, 1965; Fernandes et al., 1976; Miller & Payne, 1968; Nolen, 1972). These observations raise a fundamental issue namely what is an optimum growth rate and consequently what is an optimum diet ?

Although various life-span studies have been carried out involving restriction in rats and mice most have employed diets of fixed composition with a fixed ratio of protein to energy. It is, however, evident that the actual nutritional needs of animals may vary considerably throughout their life-span.

We report here our investigations of food restriction implemented during different phases of life. We looked at the effect during pregnancy, lactation and consequent post-weaning growth in rats as well as the effects of nutritional restriction in mice implemented well after maturation. The restriction aimed at differentiating between overall food restriction and between actual protein and/or energy restriction. It is evident, however, that these two situations cannot be entirely separated as protein can function both as an energy source and as a nutrient. Its efficiency of utilisation is in practice less influenced by its amino acid composition than by the energy value of the diet (Miller, 1973; Payne, 1972). In cases of caloric deficiency the protein will be oxidised as a caloric source while addition of an energy supply will liberate the protein for other uses.

Furthermore, classical nutritional experiments have demonstrated that when diets of constant protein concentration are fed ad libitum to rats the amount of food consumed is influenced by both protein concentration and quality. A quantitative expression of these relationships in the context of protein and/or energy restriction was extensively studied by Brody (1945) and Kleiber (1961).

A further point to be remembered in the interpretation of such experiments is that since dietary protein level much influences voluntary food intake, it is clear that some degree of anorexia will be induced by the feeding of low-protein diets. The variable magnitude of this effect depends both on the diet composition and the age of the animal (Payne, 1975; Waterlow, 1968). The food energy intake depression that results with low-protein diets is much smaller in older than in young rats.

EXPERIMENTAL AND RESULTS

Caloric requirement determining intake

We investigated caloric intake as a function of requirement in Sprague-Dawley rats of both sexes supplied ad libitum a complete diet as well as free access to either a 0, 5 or 10% sucrose solution. Starting at 5 weeks of age, for 8 weeks they were given lab chow, a 22% casein/corn starch diet or a 22% casein/50% sucrose diet. Following mating the females continued on the same ad libitum ingestion schedules throughout pregnancy, lactation and 3-weeks post-weaning. Body weight gains were similar at all sucrose concentrations, but total intakes were higher in rats supplied lab chow than those fed semi-purified diets (Fig. 1). With increasing sucrose load, decreasing voluntary intakes of dietary metabolizable energy (M.E.) and protein were noted. Thus dietary M.E. intakes by the 5 and 10% sucrose solution ingesting groups were as a per cent of the basal intake by the 0% sucrose group, respectively 72 and 47, 71 and 51, and 87 and 77, during pre-pregnancy, pregnancy and lactation. The control rats receiving the 0% sucrose solution had a Pe (protein energy as a per cent of total ingested M.E.) intake of 22.2%. With concurrent access to the 5 or 10% sucrose solution their respective intake of Pe was 15.8 or 11.5% during both pre-pregnancy and pregnancy periods. During lactation, however, these values increased respectively to 19.9% and 17.8%. Fertility and gestation indices

for rats fed a given diet were independent of the level of sucrose solution
provided.

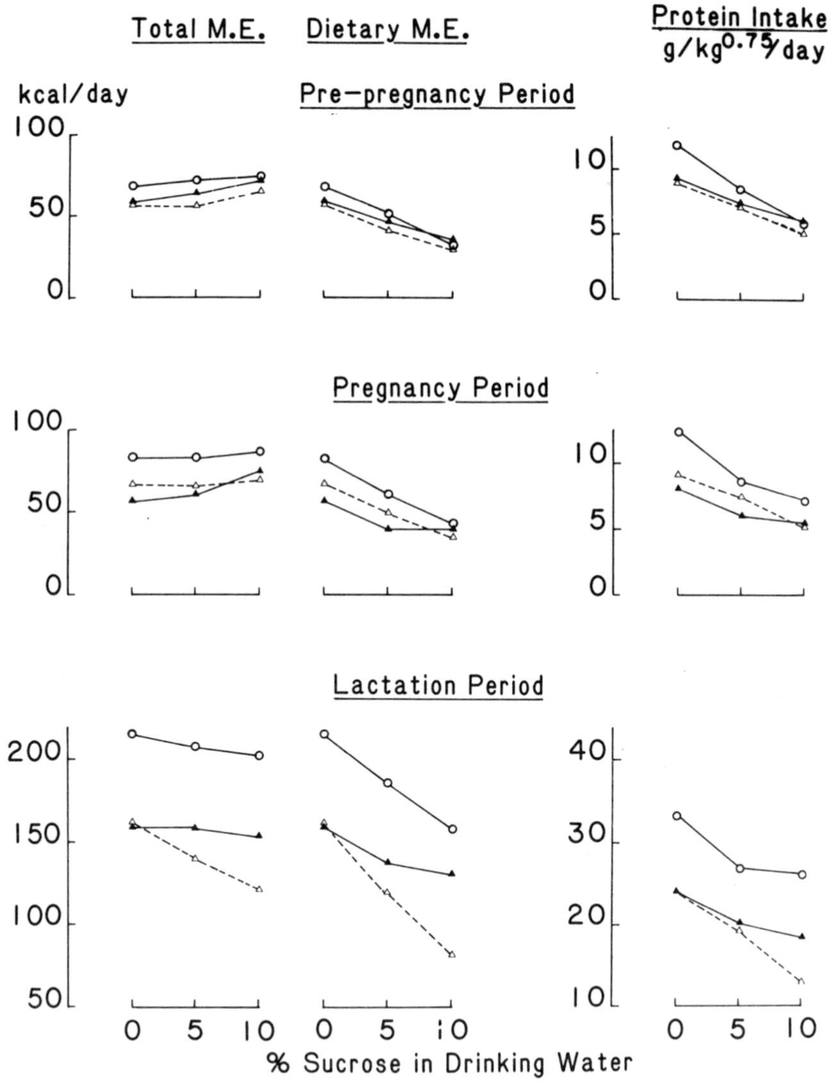

Fig. 1 : Average daily voluntary intake of total metabolizable energy
(M.E.), dietary M.E. and dietary protein by female Sprague-Dawley rats with
free access to 0, 5 or 10% sucrose solution in addition to their respective
diets (n = 6 rats in each subgroup). ---o--- lab chow; ---△--- casein-corn
starch diet; ---▲--- casein-sucrose diet.

Two conclusions may be drawn namely that a Pe intake of 11.5% supporting gestation signifies little loss of ingested nitrogen. During lactation which imposes a major demand for nutrients on the maternal organism the rat's natural food intake regulatory mechanisms prevailed over the sucrose stimulus provided.

Selective restriction of protein and/or energy

Because of the problem of discriminating between the effects of protein and energy restriction, self-selection studies in rats have been suggested as guides to assess nutritional needs in pregnancy and lactation (Collier et al., 1969; Leshner et al., 1972; Richter & Barelare, 1938; Richter et al., 1938). The separate feeding procedure originally developed by Rerat et al. (1963) was assessed by Pol and den Hartog (1966) to evaluate protein quality and efficiency of protein utilisation. The approach providing a choice of 2 diets, both ad libitum was extensively used by Ashley and Anderson (1975) and by Musten et al. (1974) to study the regulation of energy and protein intake by rats. Their experimental model, essentially involving self-selection from 2 diets with varying protein contents, has demonstrated that protein and energy intakes are independently regulated.

We adapted the separate feeding procedure of Rerat et al. (1963) to investigate the energy and protein intake regulatory response in rats subjected to selective protein or energy restriction during pregnancy and lactation. In addition, the same experimental approach was employed to study the regulatory responses and their effect on survival of mice subjected to restriction from their second year of life.

Restriction in rats during pregnancy and lactation

Five groups of 13 week-old female Sprague-Dawley rats were supplied different diets from the first day of gestation. They received either a 22% protein (high) diet ad libitum (A), a 13% protein (low) composite diet (B), or the 22% protein diet restricted to 70% of ad libitum intake (C). A further two groups received two diets (76% and 0% protein) such that one group was restricted to the protein intake of group C but with energy supplied <u>ad</u> <u>libitum</u> (D) while the other group was restricted to the energy intake of group C but with protein supplied <u>ad</u> <u>libitum</u> (E).

The different restrictions variably influenced gestational and lactational performance of the dams. However, none of the restriction types seriously affected reproductive performance.

In the groups subjected to selective protein or energy restriction the rats, throughout pregnancy and lactation, regulated distinctly their voluntary ingestion of energy (0% protein diet) or of protein 76% protein diet). On protein restriction, the amount chosen of the 0% protein diet was such, that average Pe intake was 18% whilst on energy restriction, they consumed a large amount of the 76% protein diet so that average Pe selected amounted to 44% (Table I, Fig. 2).

Table I : Selective protein or non-protein energy restriction in female rats during gestation and lactation (Values are means)

	Diet treatments and restrictions				
	A	B	C	D	E
Dams (n)					
Gestational gain, g	39.1	26.3	17.3**	36.4	6.1**
Gestational intake, kJ/d	304	285***	217***	283***	308
Lactational intake, kJ/d	720	471***	451***	592***	677**
Pe - gestation, %	21.8	13.0	21.8	18.7	42.3
Pe - lactation, %				18.0	44.7
Body fat at end of lactation, %	10.2	9.1	4.3***	10.4	6.0*
Young rats (n)					
Birth weight, g	6.1	5.5**	6.1	5.5**	5.9
Weaning weight, g	58.0	34.6***	40.2***	45.6***	42.9***
Body fat (weaning), %	15.8	16.7	10.7***	17.1	11.8***

*, **, *** are significance at $p < 0.05$, 0.01 and 0.001 compared to treatment A (22% casein protein).

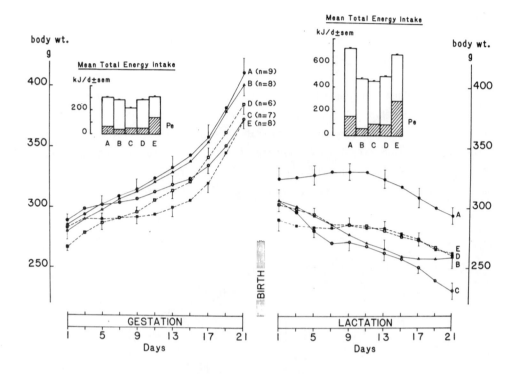

Fig. 2 : Changes in body weight and average daily energy and protein-energy (Pe) intake of female Sprague-Dawley rats during gestation and lactation.

Carcass analysis revealed that animals of groups B, C, D and E laid down less fat reserves during gestation compared to reference group A, while animals of groups B, C and D mobilised this fat more rapidly during lactation. Animals in group E (selective energy restriction) despite the very small reserves (non-foetal gain = 6.1 g) met lactational costs through their high Pe intake (44%) compared with an average Pe intake of 18% by animals (group D) subjected to selective protein restriction.

The low body fat in both dams and weanlings of group D and E suggests that the development of maternal fat reserves was either inadequate or that their mobilisation commenced already early in lactation.

Restriction in mice after maturity

Male and female mice which received a standard lab chow until 58 weeks of age were allotted to five treatment groups. One group continued to receive the lab chow ad libitum (A) and another group received a 22% protein (lactalbumin) diet ad libitum (B). A further group received the 22% protein (lactalbumin) diet but restricted to 70% of the ad libitum intake level (C). Groups D and E were subjected to selective restriction of protein or non-protein energy (to achieve the levels of intakes of mice in group C) through restricted feeding of a 66% protein (lactalbumin) diet and ad libitum offering of a 0% protein diet and restriction of the 0% protein diet and free access to the 66% protein (lactalbumin) diet respectively.

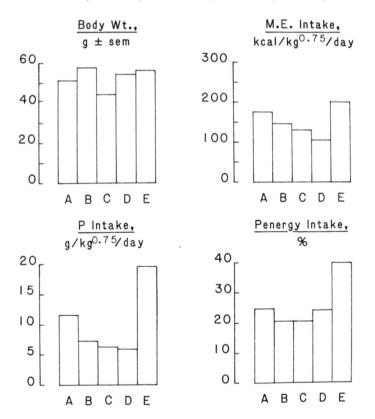

Fig. 3 : Average body weight, and daily intake of M.E., protein and protein energy expressed as a per cent of total M.E. by male mice subjected to different diet treatments and restrictions between 58 and 120 weeks of age.

Average body weights (between weeks 58 and 120) were similar whether protein or energy was restricted (groups D and E); they were also comparable to those of mice ingesting the lab chow <u>ad libitum</u> or the 22% protein (lactalbumin) diet (groups B and A). With the overall restriction (group C), the mice rapidly lost 20% of their body weight. The ingestion of the 0% or the 66% protein diet on selective protein or energy restriction was such that Pe intake was 23% and 41% respectively (Fig. 3). These observations in ageing mice are strikingly similar to the corresponding values of 18% and 44% obtained in pregnant and lactating rats.

Of all the treatments, the restricted feeding of the 22% protein diet (C) promoted the longest 50% survival time and the highest survival rate at 120 weeks (Fig. 4). Furthermore, the survival rate of mice on selective energy restriction with a Pe intake of 41% was higher than that of mice on protein restriction but with a Pe intake of 23%.

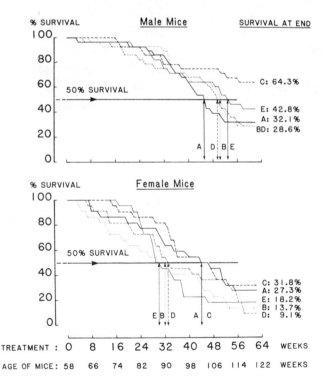

Fig. 4 : Survival curves for male and female mice on different diet treatments commenced at 58 weeks and monitored till 120 weeks of age.

DISCUSSION

Studies of the total caloric intake of ad libitum fed rats offered both a sucrose solution and lab chow concurrently (Jacobs & Sharma, 1969; Kanarek & Hirsch, 1977) revealed that at all the concentrations of sucrose solution except the lowest (3%) the animals maintained the proportion of their caloric intake from sucrose at a fixed level. Recently several authors have reported work on overeating and obesity in rats under conditions of self-selection of dietary components (Kratz & Levitsky, 1979) or overeating and diet selection patterns (Hirsch et al., 1982; Hirsch & Walsh, 1982) in either case with concurrent free access to sucrose solutions. The basis for such studies is that stimulation induced eating and drinking, if rewarding can overcome natural food intake regulatory mechanisms. The animals would then eat in response to the stimulus (e.g. sugar solution) and thereby become hyperphagic and eventually obese.

In another study Hartsook et al. (1973) investigated the energy metabolism of growing rats as influenced by dietary protein (casein) content and ratio of fat to carbohydrate calories. They observed optimal energy gain, maximum heat of basal metabolism and maximal body nitrogen gain at respectively 37, 39 and 44% Pe level in the diet. The inference being the "food intake is regulated by energy requirement" which itself is a function of body weight and of the quality and quantity of protein.

On the other hand, animals given the choice between two diets of different protein levels showed Pe intakes characteristic for a particular protein source : 43% for gluten but only 33% for casein (Musten et al., 1974). Pol and den Hartog (1966) found that weanling rats allowed to self select from 60% and 0% protein diets showed a higher Pe intake from gluten (poor protein) than the better quality potato protein. Again Ashley and Anderson (1975) investigating the effects of additions of the most limiting amino acids to gluten, casein and zein on the self-selection of protein and energy by weanling rats given a choice of two diets differing only in protein level concluded that the selection of protein and energy by the weanling rat is not related to the nutritional quality of the protein fed.

However, in our studies the increased needs for energy during pregnancy and lactation are likely to have contributed to the high Pe (44%) selection by the rats subjected to selective energy restriction. Functional cycles of energy and protein conservation and release in dams are known to operate normally even under marginal dietary restriction (Blaxter, 1964; Fell, 1972; Naismith, 1969, 1981; Naismith & Morgan, 1976).

Although undoubtedly food restriction in rodents delays the emergence of diseases which shorten the maximal life-span, little work has so far been done to determine the usefulness of diet restriction in mature life on length of life. The little that has been done is highly controversial. Barrows and Roeder (1965) reported that food restriction applied after 1 year of age failed to increase life-span. In contrast, Miller and Payne (1968) observed that rats fed a stock diet until 100 days of age and then transferred to a 4% protein diet showed little further increase in their body weight and lived 980 days as compared with the 760 days of their ad libitum stock fed diet controls. Stuchlikova et al. (1975) also found that food restriction applied to rats at 1 year of age can still increase longevity. Our own results which showed the highest survival rate in mice undergoing 30% restriction as well as better survival for mice on selective energy restriction than on selective protein restriction are in agreement with the latter workers.

CONCLUSIONS

Since energy metabolism and protein metabolism are intimately linked, attempts to completely dissociate the effects of one from the other are difficult to implement. Nevertheless, the present work, in agreement with previously published reports in weanling rats, demonstrates that energy and protein intakes are independently regulated even in the pregnant and lactating rat and the adult mouse.

Although the issue of dietary restriction in adult animals and its effects on life-span are still controversial, our data in mice restricted only in the second year of their life supports the view of other authors that such restriction may still promote the highest survival.

Furthermore, although restriction of calories exerts strong effects as shown by our study on pregnant and lactating rats, yet the females were able to compensate the lactational costs by increasing their Pe intake.

The selective restriction procedure must be considered as a suitable model for such studies, however, further studies are necessary to fully investigate the effects of protein and energy restriction starting at gestation and continuing into old age.

REFERENCES

Ashley, D.V.M. & Anderson, G.H. (1975) Effects of the most limiting amino acids of gluten, casein and zein on the self-selection of protein and energy. J. Nutr. 105:1405-1411.

Barrows, C.H. & Roeder, L.M. (1965) The effect of reduced dietary intake on enzymatic activities and life span of rats. J. Geront. 20:69-71.

Berg, B.N. & Simms, H.S. (1960) Nutrition and longevity in the rat. II. Longevity and onset of disease with different levels of food intake. J. Nutr. 71:255-264.

Blaxter, K.L. (1964) Protein metabolism and requirements in pregnancy and lactation. In : Mammalian Protein Metabolism. Vol. II (Ed. H.N. Munro, J.B. Allison) Academic Press, New York & London, pp. 173-223.

Brody, S. (1945) Metabolism and pulmonary ventilation in relation to body weight during growth. In : Bioenergetics and Growth. Hafner Press & Collier-Macmillan, New York & London, pp. 404-469.

Collier, G., Leshner, A.I. & Squibb, R.L. (1969) Self-selection of natural and purified protein. Physiol. Behav. 4:83-86.

Fell, B.F. (1972) Adaptations of the digestive tract during reproduction in the mammal. Wrld Rev. Nutr. Dietet. 14.180-256.

Fernandes, G., Yunis, E.J. & Good, R.A. (1976) Influence of diets on survival in mice. Proc. Natl. Acad. Sci. USA 73:1279-1283.

Hartsook, E.W., Hershberger, T.V. & Nee, J.C.M. (1973) Effects of dietary protein content and ratio of fat to carbohydrate calories on energy metabolism and body composition of growing rats. J. Nutr. 103:167-178.

Hirsch, E., Dubose, C. & Jacobs, H.L. (1972) Overeating, dietary patterns and sucrose intake in growing rats. Physiol. Behav. 28:819-828.

Hirsch, E. & Walsch, M. (1982) Effect of limited access to sucrose on overeating and patterns of feeding. Physiol. Behav. 29:129-134.

Jacobs, H.L. & Sharma, K.N. (1969) Taste versus calories : Sensory and metabolic signals in the control of food intake. In : Neural Regulation of Food and Water Intake (Ed. P.J. Morgane) Ann. N.Y. Acad. Sci. 157:1049-1061.

Kanarek, R.B. & Hirsch, E. (1977) Dietary-induced overeating in experimental animals. Fed. Proc. 36:154-158.

Kennedy, G.C. & Pearce, W.M. (1958) The relation between liver growth and somatic growth in the rat. J. Endocrinol. 17:149-157.

Kleiber, M. (1961) The Fire of Life : An Introduction to Animal Energetics. John Wiley, New York & London, 454 p.

Kratz, C.M. & Levitsky, D.A. (1979) Dietary obesity : Differential effects with self-selection and composite diet feeding techniques. Physiol. Behav. 22:245-249.

Leshner, A.I., Siegel, H.I. & Collier, G. (1972) Dietary selection by pregnant and lactating rats. Physiol. Behav. 8:151-154.

McCay, C.M., Maynard, L.A., Sperling, G. & Barnes, L.L. (1939) Retarded growth, life span, ultimate body size and age changes in the albino rat after feeding diets restricted in calories. J. Nutr. 18:1-13.

McCay, C.M., Sperling, G. & Barnes, L.L. (1943) Growth, ageing, chronic diseases and life span in rats. Arch. Biochem. 2:469-479.

Miller, D.S. (1973) Protein-energy interrelationships. In : Proteins in Human Nutrition (Ed. J.W.G. Porter, B.A. Rolls) Academic Press, London & New York, pp. 93-101.

Miller, D.S. & Payne, P.R. (1968) Longevity and protein intake. Exp. Geront. 3:231-234.

Musten, B., Peace, D. & Anderson, G.H. (1974) Food intake regulation in the weanling rat : Self-selection of protein and energy. J. Nutr. 104:563-572.

Naismith, D.J. (1969) The foetus as a parasite. Proc. Nutr. Soc. 28:25-31.

Naismith, D.J. (1981) Diet during pregnancy - a rationale for prescription. In : Maternal Nutrition in Pregnancy - Eating for Two ? (Ed. J. Dobbing) Academic Press, London, pp. 21-40.

Naismith, D.J. & Morgan, B.L.G. (1976) The biphasic nature of protein metabolism during pregnancy in the rat. Br. J. Nutr. 36:563-566.

Nolen, G.A. (1972) Effect of various restricted dietary regimens on the growth, health, and longevity of albino rats. J. Nutr. 102:1477-1494.

Payne, P.R. (1972) Protein quality of diets, chemical scores and amino acid imbalances. In : Protein and Amino Acid Functions (Ed. E.J. Bigwood). International Encyclopaedia of Food and Nutrition. Vol. II. Pergamon Press, London & New York, pp. 259-306.

Payne, P.R. (1975) Influence of energy intake on ageing and longevity. In : Proc. 9th Int. Congr. Nutr., Mexico 1972. Vol. 1. Karger, Basel, pp. 353-361.

Pol, G. & den Hartog, C. (1966) The dependence on protein quality of the protein to calorie ratio in a freely selected diet and the usefulness of giving protein and calories separately in protein evaluation experiments. Br. J. Nutr. 20:649-661.

Rerat, A., Henry, Y. & Jacquot, R. (1963) Relation entre la consommation spontanée d'énergie et la rétention azotée chez le rat en croissance. C.R. Acad. Sci. Paris 256:787-789.

Richter, C.P. & Barelare, B., Jr. (1938) Nutritional requirements of pregnant and lactating rats studied by the self-selection method. Endocrinology 23:15-24.

Richter, C.P., Holt, C.E. & Barelare, B., Jr. (1938) Nutritional requirements for normal growth and reproduction in rats studied by the self-selection method. Am. J. Physiol. 122:734-744.

Ross, M.H. (1959) Protein, calories and life expectancy. Fed. Proc. 18: 1190-1207.

Ross, M.H., Lustbader, E. & Bras, G. (1976) Dietary practices and growth responses as predictors of longevity. Nature 262:548-553.

Silberberg, M. & Silberberg, R. (1955) Diet and life span. Physiol. Rev. 35:347-362.

Stuchlikova, E., Juricova-Horakova, M. & Deyl, Z. (1975) New aspects of the dietary effect of life prolongation in rodents. What is the role of obesity in aging ? Exp. Geront. 10:141-144.

Waterlow, J.C. (1968) Observations on the mechanim of adaptation to low protein intakes. Lancet ii:1091-1097.

Widdowson, E.M. & Kennedy, G.C. (1962) Rate of growth, mature weight and life span. Proc. Roy. Soc. London B156:96-108.

BEHAVIOURAL STRATEGIES IN THE REGULATION OF FOOD CHOICE

Peter D. LEATHWOOD and David V.M. ASHLEY

Nestlé Products Technical Assistance Co. Ltd, Research Department,
CH-1814 La Tour-de-Peilz, Switzerland

SUMMARY

The maintenance of nutrient and energy balance in the body depends on both metabolic and behavioural mechanisms, and is integrated by the brain. The regulatory system was developed by natural selection and not by mechanical engineers. Thus, rather than having unitary mechanisms regulating intake of each nutrient, evolution has incorporated and used a multitude of behavioural traits and metabolic adaptations. The criterion for inclusion was that each one conferred a persisting advantage in the prevailing environment.

Behavioural strategies in food choice include : innate preference for sweetness and an aversion towards bitter tastes, a hesitancy towards unknown foods, preference for variety among familiar foods, and a special ability (long delay learning) to acquire information about both positive and negative metabolic consequences of eating different foods. In man, these more basic mechanisms interact with and are complemented by cognitive, social and cultural influences on food choice. In a very few cases, such as regulation of energy, sodium and (perhaps) protein intakes, feeding behaviour is also guided by signals from specific internal receptors. However, for most nutrients, appetites seem to be non-specific and learned.

Using studies on the regulation of protein intake from our own and other laboratories as examples, this review illustrates how innate preferences, learning, social interactions, metabolic adaptation and diet-induced changes in brain neurotransmitter metabolism can all play a role in subjective decisions about what to eat.

* * *

Maintaining an adequate supply of energy and nutrients to the body is the first priority of animal life. Only when needs for growth and tissue maintenance have been satisfied can the individual even begin to indulge in sexual behaviour and so assure the survival of his or her genes. This overall nutritional balance depends on metabolic and behavioural mechanisms which are ultimately regulated and integrated by the brain, and it is important to remember that the system was designed by natural selection, not by mechanical engineers. Thus, rather than producing unitary mechanisms regulating the intake of each nutrient, evolution has incorporated and used a multitude of behavioural traits and metabolic adaptations. The criterion for inclusion was that each one conferred a persisting advantage in the prevailing environment, so it should not be surprising that, according to the conditions in which an organism is studied, different regulatory mechanisms are called into play, or even that, under extreme circumstances, regulation breaks down entirely.

The multiplicity of different mechanisms influencing food choice and food intake has led to two distinct types of scientific enquiry. The first aims at demonstrating that a particular factor - a nutrient, a drug, a neurotransmitter or a part of the brain - can influence some aspect of food intake or food choice under a specific, and often extreme, set of experimental conditions. Such studies can be instructive but it is often difficult to know to what extent the results from them can be generalised. The second approach aims at understanding how different mechanisms are integrated and applied under different circumstances, and at demarcating the limits of application of each one.

This chapter aims to concentrate on the latter approach. First, we will show how regulation of food choice fits into the larger context of regulation of nutrient balance. We will then consider how metabolic mechanisms offset the need for a very precise regulation of intake for each nutrient, and will discuss the different behavioural strategies which evolution has developed to help maintain adequate levels of nutrients in the body. Finally, using results from our own laboratory, we will illustrate how these different strategies operate on protein selection to ensure amino acid availability.

NUTRIENT BALANCE

The balance for any nutrient or for energy in the body can be expressed as follows :

$$\text{Intake} = \begin{array}{c} \text{metabolic} \\ \text{requirements} \end{array} + \begin{array}{c} \text{losses and excretion} \\ \text{of excesses} \end{array} + \text{storage}$$

Equilibrium may be maintained by compensatory modification of any or all terms in the equation, so, as Le Magnen (1971) has pointed out, studying any one of the above aspects of energy or nutrient regulation in the body whilst ignoring the others, can be very misleading.

There is an abundance of evidence showing how metabolic adaptation contributes towards maintaining nutrient balance. If intake of a nutrient increases, excesses can be stored or excreted. When intake is low, stores can be depleted and losses due to excretion may decrease.

The efficiency of these mechanisms of metabolic adaptation means that there is no need for extremely precise regulation of intake for each nutrient. On the other hand, when rats are allowed to select from several foods their patterns of selection are usually quite stable. Furthermore, selection adjusts to changing physiological need, as is shown by the changes in nutrient intake of pregnant or lactating rats (Leshner et al., 1972; Richter & Barelare, 1938). Similarly, when metabolic adaptation is compromised, behaviour (i.e. strategies of food selection) can usually cope with the problem of maintaining balance. A good example of this phenomenon is the increase in drinking from a dilute salt solution by adrenalectomised rats (Richter, 1936; for other examples see Rozin, 1976).

BEHAVIOURAL STRATEGIES

The strategies and mechanisms which are involved in behaviour's contribution to maintaining nutrient balance include : innate preferences and aversions; innate behaviour patterns such as hesitancy towards new foods, a preference for familiar foods; a preference for variety among familiar foods; true specific appetites (which exist for a few nutrients); special abilities (such as specific association and long delay learning) to acquire information about both positive and negative metabolic consequences of eating different foods; social influences on food choice. For man, these mechanisms are the basis upon which cognitive and cultural influences on food choice are built (Fischler, 1981).

No single strategy gives a complete explanation of observed patterns of food choice, but, in a given situation, one or more may be called into play.

Innate preferences and aversions

The geometry of the head, with the special sense organs grouped near the mouth, ensures that potential foods are carefully examined before being consumed. The sense of taste is intimately involved in both innate and learned preferences and aversions, so even when in the mouth, it is not too late for the final decision as to whether a food should be rejected or accepted (Rozin & Fallon, 1981).

Innate preferences for sweet foods such as ripe fruits have evident survival value for the omnivore in that they often provide a safe and quick source of calories (Rozin, 1976). Similarly, innate aversion to bitter tastes (Steiner, 1973) may also owe its existence to the selective advantage enjoyed by animals which spontaneously rejected the (often poisonous) bitter alkaloids which are widely distributed in the plant kingdom.

These innate preferences are not fixed and can be changed by experience so that omnivores can learn to avoid sweet tastes if the latter are followed by unpleasant metabolic consequences (Arimanana & Leathwood, unpublished; Garcia et al., 1974), or will come to prefer bitter tasting foods that are associated with positive metabolic effects (Kratz et al., 1978).

Innate patterns of behaviour

Just as physical traits such as specific receptors for sweet or bitter were subject to natural selection, the same process has occurred with behaviours, so that some behaviour patterns which appear intelligent in fact arise from phylogenetic memory rather than from learning. In rats and men an important group of behavioural strategies in feeding belong to this category.

First there is the approach to new foodstuffs. On one hand, food sampling is essential in adjusting to new situations and maintaining intake from a wide variety of food sources but, on the other, any potential new food could be poisonous. Thus omnivores exhibit ambivalent behaviour towards new foods. They are both hesitant and curious. Wild rats approach a new food with great caution, taste it, and then wait some time before consuming any more (Rzoska, 1953). This strategy increases their chances of identifying poisonous foods without killing themselves in the process of doing so (see next section). Once, however, a new food has been tasted and does no harm, it becomes a safe familiar food and is no longer approached hesitantly (Rozin, 1976).

Another important innate behavioural strategy is preference for variety within the domain of familiar foods. Rats offered a choice of nutritionally adequate diets tend to eat from several of them rather than just one; this preference for variety is so powerful that rats will even work to obtain a diet which is readily available from an alternative source (Leathwood et al., 1981) rather than eat all their food from the one source. Similarly, in man, the attractiveness of a favourite dish palls rapidly if it is presented at each mealtime.

These relatively simple behaviour traits have such self-evident survival value that it is easy to imagine how by natural selection they have become part of the phylogenetic memory of instinctive behaviour.

Poison avoidance

As pointed out above, a major feeding problem for omnivores is that among the multitude of potential foods some may be poisonous. Innate taste aversions and behaviour patterns such as the "taste and wait" (Rzoska, 1953) approach to new foods are helpful in avoiding poisons, but the identity of each new poison must still be learned. As might be expected, several different strategies for poison avoidance are available. Social factors probably play an important role, so that many young animals will not eat a food their mother has learned to avoid (Wyrwicka, 1976). Individual rats also show special learning abilities well adapted to the special problems of poison avoidance.

They make specific associations between the tastes of new foods and the metabolic consequences of having eaten them (Garcia et al., 1974). If made ill by X-irradiation or by lithium chloride injection shortly after ingesting a novel flavoured food, rats will thereafter avoid eating foods with that flavour. In constrast, giving them an electric shock after they have consumed a novel food is a much less effective aversion therapy (Garcia et al., 1974). Thus an intestinal upset or malaise is specifically associated with the taste of a recently eaten food, but the pain from an electric shock - which the rat can quickly learn to associate with a sound or a light stimulus (Leathwood, 1978) - is not associated with a taste.

Added to this is long delay learning - the ability to associate a conditioned stimulus (the taste of a new food) with an unconditioned stimulus (malaise), even when the two are well-separated in time. If rats are made ill 6-12 hours after ingesting a novel-flavoured diet, they will subsequently avoid that flavour. This learning occurs in a single trial, but if, subsequently, rats are offered only the flavoured diet and they become hungry enough to taste it again, extinction of the aversion is also rapid (Rozin, 1976).

The combination of hesitancy towards new foods and an ability to make specific associations between their sensory characteristics and their unpleasant metabolic effects, even if the latter occur after some time has passed, offers a quite satisfactory explanation as to how rats avoid poisoning themselves. It does not, however, explain how they come to choose an adequate nutrient selection from the non-poisonous food sources that remain.

True specific appetites

For some nutrients, appetite is also guided by specific receptor systems. For example, sodium ions are detected by the sense of taste and when rats are sodium-deficient, they show an immediate preference for salt solutions. This preference occurs before any metabolic repletion has had the time to occur (Denton, 1982; Richter, 1936). Similarly, for water (if this can be considered a nutrient) there are specific receptor systems which trigger thirst sensations as the animal becomes dangerously water-depleted (Rolls & Rolls, 1981). This physiological mechanism is seldom used because most normal drinking is "prophylactic" in that animals and men will spontaneously drink well before depletion-driven thirst signals begin to operate. Control of energy intake depends on the integration by the brain of a complex of innate and learned behaviours some of which are linked to internal detectors (Booth, 1981; Cioffi et al., 1981).

But, for most nutrients, the evidence available suggests that appetites for them are acquired without the aid of specific receptor systems (Rozin, 1976).

Acquiring specific appetites

One further manner in which rats learn to prefer diets which satisfy the need for a particular nutrient is illustrated by Rozin's analysis (1967) of the feeding strategies employed by thiamine-deficient rats.

After several days on a thiamine-deficient diet, rats begin to eat less food and lose weight. If they are offered a choice between the thiamine-deficient diet and a new thiamine-rich diet, they will eat the latter almost exclusively for several days (Harris et al., 1933). However, they will do the same even if they are given thiamine injections before being offered the new diet (Rozin, 1967) suggesting that they choose the new diet because the original one has become aversive rather than because they recognise thiamine in the new diet.

Confirmation of this view comes from two studies described in Rozin (1967). The first experiment (Scott & Vernay, 1947) confirms that thiamine-deficient rats do not have specific thiamine detectors to help them choose the right food. Rats were first made thiamine-deficient, then offered a new thiamine-rich diet with a distinctive flavour (anise). The rats ate it. They were then offered a choice of diets containing anise without thiamine or thiamine without anise. The rats followed the anise. This shows that the rats learned to associate recovery from deficiency with the dominant sensory characteristic of the diet. They did not detect thiamine per se.

In the second study, rats made thiamine-deficient (on what we might call diet A) were offered a choice between diet A and new diet B. For some animals, thiamine was added to A, for others, to B. All thiamine-deficient animals chose diet B (Fig. 1). If B contained thiamine they continued to eat it. If A contained thiamine, the rats persisted with B for a few days then abruptly changed their choice (Rodgers & Rozin, 1966). This suggests that for deficient rats any new food is preferred (i.e. they show an exaggerated preference for novelty) but will continue to be preferred only if it alleviates the deficiency.

We do not know what metabolic signal directs the deficient rats to prefer the thiamine-rich diet once they have tasted it, but some observations in

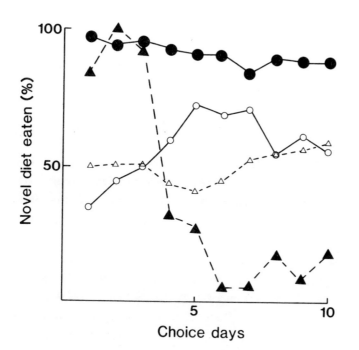

Fig. 1 : Preference for a novel diet by thiamine-deficient rats : Control or thiamine-deficient rats were offered a choice between the original diet and a novel diet; for some animals, thiamine was added to the novel diet, for others to the original diet. Each point represents the percent of food eaten from the novel diet by : A, deficient rats, thiamine in the novel diet ●——●,; B, control rats, thiamine in the novel diet o——o; C, deficient rats, thiamine added to the original diet ▲---▲; D, control rats, thiamine added to original diet Δ---Δ. Each group contained 3 rats (redrawn, with permission, from Rozin, 1967).

man offer a possible explanation and allow a tentative extrapolation of these results to learned appetites for other nutrients. Thiamine-deficient men report feeling irritable and depressed; thiamine supplements bring about a rapid improvement in mood. If rats also feel "better" soon after eating one of the diets, they may associate the change in metabolic (or mental) state with the taste, texture and placement of the new diet (i.e. will exhibit specific association and long delay learning). This, coupled with innate tendencies to prefer new foods and to taste them one at a time, should quickly lead to preference for the appropriate diet.

The development of specific appetites for other vitamins and minerals have been less completely studied. Most of them seem to be acquired, with varying degrees of success, in a similar manner to that described for thiamine (see Rozin, 1976 for a review). It has been suggested that important factors in the success in learning to acquire the appetite are the rapidity and character of alleviation of deficiency symptoms following ingestion of the new, adequate foodstuff (Leathwood & Ashley, 1981).

In summary, rats are provided with innate behaviour patterns (curiosity towards new foods, preference for variety among familiar foods) which lead them to sample many different foodstuffs. This increases the probability that they will ingest an adequate amount of each essential nutrient. When deficient, they have more strategies (exaggerated preference for novelty; waiting several hours between tasting each new food), and special learning characteristics (specific association; long delay learning), which favour acquisition of specific appetites for foods which alleviate a deficit without recourse to specific receptors for each nutrient. In a few cases preferences can be guided by specific internal receptor systems but these seem to function more as fail-safe devices to help prevent serious imbalances occurring rather than as controls on normal eating and selection on a meal-to-meal or mouthful-to-mouthful basis (Booth, 1981).

SOCIAL AND CULTURAL INFLUENCES ON FOOD CHOICE

Social and cultural influences are very important determinants of food selection and nutrient intake, particularly in normal eating. Even among rats there is strong evidence for the importance of social transmission in the avoidance of poisons (Wyrwicka, 1976) and in preferences among nutritionally adequate foods (Galef, 1976; Leprohon & Anderson, 1982). Furthermore, group-housed rats offered a multiple food choice (cafeteria) are consistently more successful in developing adequate choice patterns than are individually housed animals (Lat, 1967).

In man, culture provides the rules as to what should be eaten or not. These rules are so coercive that, if one wanted to know to what culture a person belonged and one only had the right to pose one question about that person, the best question might well be "What dish do you usually eat for dinner ?". Very often these rules encode intelligent nutrition practices (for example, the inclusion of lime in the preparation of tortillas releases niacin - Kodicek & Wilson, 1959), although other cultural food practices seem to be illogical or counter-productive (for examples see : de Garine, 1979). However, the social anthropology of food choice is outside the terms of reference of this chapter and the interested reader is referred to the following reviews for a more detailed treatment of this topic : Cohen, 1977; Fischler, 1981; Solms & Hall, 1981.

PROTEIN INTAKE

Animals and men must eat protein to maintain an adequate supply of amino acids to the body. In this section, we will show how, even in very simple choice situations, metabolic adaptation and many of the behavioural strategies outlined above are involved in achieving this end. Furthermore, we will also try to evaluate the extent to which central serotoninergic mechanisms are involved in influencing protein selection.

Metabolic adaptation

Metabolic adaptation can cope with a wide range of protein intakes and rats grow perfectly well on diets containing anything from 15 to 55% of a good quality protein. If intake is in excess of needs, amino acids can be transaminated or deaminated, the nitrogen is excreted as ammonia or urea and the carbon skeleton incorporated into energy metabolism. In periods of deficiency, metabolism adapts to limit amino acid loss (Das & Waterlow, 1974; Krebs, 1964; Munro, 1964; Oddoye & Margen, 1979) and stores are depleted.

On the other hand, it has been suggested (Musten et al., 1974) that, if given the choice, rats will select quite stable levels of protein.

There is no self-evident advantage in tight regulation of protein selection although : (1) adaptive mechanisms to eliminate excess nitrogen involve a metabolic cost (Munro, 1964) and (2) there does appear to be a level of protein intake which leads to maximum nitrogen retention (Hartsook et al., 1973), so there may be an optimum level of protein intake.

Behavioural analysis of protein selection

The most common approach to the analysis of protein choice has been either to measure mean choice over a period of weeks (Musten et al., 1974; Peters & Harper, 1981) or to examine the effects of drugs (Wurtman & Wurtman, 1977; Blundell et al., 1979) or brain lesions on protein choice (Ashley et al., 1979). Another approach (Leathwood & Ashley, 1983) is to examine the manner in which rats offered a choice of diets come to select adequate amounts of protein.

Development of protein selection

Fig. 2 shows daily measures for body weights, food intake and protein selection by 5 adult rats offered a choice of 0% and 40% casein diets for 2 weeks. The figure shows :

(i)   All rats tasted both diets on the first day.

(ii)   One rat (*) chose an adequate amount of protein from the first day.

(iii)   The others preferred to eat most of their food from the protein-free diet.

(iv)   Those that ate inadequate amounts of protein began to lose weight and became anorectic.

(v)   After 3 to 11 days, these animals abruptly began to eat adequate amounts of protein. They immediately increased their food intake and resumed body growth.

This series of observations suggests two important conclusions :

(i)   Initial choice is based on palatability, not on long-term protein needs.

(ii)   Adequate selection can occur either by chance (i.e. because palatability coincides with physiological needs) or is learned.

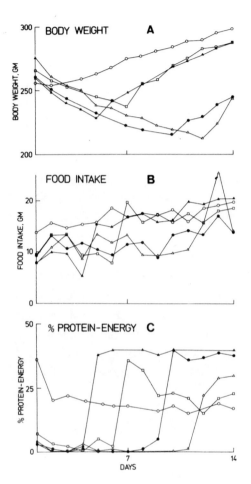

Fig. 2 : A) Body weight; B) Daily food intake; C) % P-E selection by 5 adult male rats offered a choice of 0 and 14% casein diets during 4 weeks (from Leathwood & Ashley, 1983, with permission).

These results also suggest that the way protein appetite is learned does not depend on specific detector systems and, if a deficit occurs, protein appetite is acquired in a similar fashion to thiamine appetite. Thus, if the initial choice was inadequate, the deficient diet acted like a slow poison and became progressively more aversive (this seems a reasonable assumption in that the rats lost weight and ate less food). The alternative diet eventually became relatively more attractive and was eaten. As it alleviated the deficiency, it became even more palatable, which explains the abrupt switch to high levels of protein choice.

The observations and conclusions outlined above are valid for a range of situations. If other choices were offered (e.g. 0% and 60% or 10% and 40% casein) the evolution remained basically the same. With diets made more palatable by the addition of sugar or by using meat extract as the source of protein, more animals chose an adequate amount of protein immediately. With powder diets (Hamarsten Casein is finely powdered and seems to be aversive to rats) the rats chose the protein-free diet for longer than with the same diets in granulated form (Leathwood, unpublished data). Finally, when the experiment was repeated with weanling rats, initial selection of the protein-free diet was the same, but preference for protein developed more rapidly (Leathwood & Ashley, 1983) perhaps because the weanlings became protein-deficient more quickly than adults.

In summary, the different patterns of development of protein selection in a range of experimental situations fits well with the idea that protein appetite is non-specific and acquired. However, once rats did select enough protein, choice tended to remain fairly stable (see Fig. 2). From the above studies, it was not possible to say if this stability occurred because the rats were detecting and responding to the level of protein in the diet, or if they were simply avoiding aversively low or high protein intakes and choosing on the basis of palatability.

Stability of protein selection

When rats are selecting from two diets (one containing protein; the other protein-free), and the protein-containing diet is diluted with water (Rozin, 1968) or fibre (Musten et al., 1974), they may adjust their food choices and intakes to at least partially correct for the change in protein density. This suggests that rats can respond to the level of protein in food.

We have extended these observations using an alternative approach (Leathwood & Ashley, 1983). Rats were given a pair of diets to select from and, when protein choice had stabilised, a new pair of diets was offered. The compositions of the new diet pairs were such that each rat, by appropriate adjustment of selection, could maintain a constant level of protein intake across the change in the diets.

The results of two different studies are illustrated in Figs 3 and 4. In the first study (Fig. 1), 3 groups of 8 weanling rats were offered two different diet pairs for two weeks, then the diets were changed. There was :

(i)   an initial preference for low protein diets in all three groups;

(ii)   a rapid transition to adequate protein selection;

(iii) a more or less stable mean % P-E selection in each group (although different among the 3 groups);

(iv)   large and statistically significant changes in selection by each group when the diet pairs were changed after 14 days.

We may conclude that whatever drove the weanling's selection patterns during the first period, it was not a mechanism regulating protein intake.

For adults (Fig. 4) the picture is different. Adult rats showed :

(i)   a very much slower evolution towards stable mean choice;

(ii)   very large inter-individual variability;

(iii) no significant change in mean selection when the diets changed.

The conclusions we can draw is that perhaps adult rats do detect and react to protein level or protein-carbohydrate balance in the diet.

Brain serotonin metabolism and protein selection

It has been suggested that the concentration of 5-hydroxytryptamine (5-HT, serotonin) in the brain may signal the current level of protein (or carbohydrate) intake and so influence subsequent protein choice (Anderson et al., 1982; Leathwood & Ashley, 1981; Wurtman et al., 1981).

The evidence (still incomplete and contradictory) for this hypothesis is as follows :

a) Brain tryptophan (TRP) levels can influence brain 5-HT synthesis : 5-HT in the brain is synthesized from brain TRP. As the Km (Michaelis constant) of tryptophan hydroxylase for TRP is higher than the normal concentration range of TRP in the brain, the rate of 5-HT synthesis may depend on brain TRP levels (Wurtman et al., 1981).

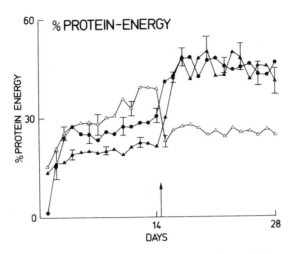

Fig. 3 : Mean % P-E selection by 3 groups of weanling rats offered the following amounts of casein. Group 1 ●——● : period 1 (days 1-14) 0% and 40% casein diets; period 2 (days 15-28), 20% and 60%. Group 2 △——△ : period 1, 0% and 60%; period 2, 10% and 40%. Group 3 ▲——▲ : period 1, 10% and 40%; period 2, 0% and 60% casein diets. The change in diets offered is marked with an arrow.

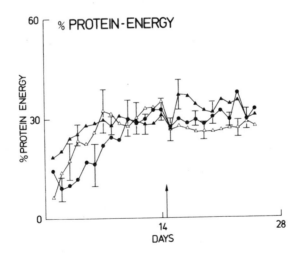

Fig. 4 : Mean % P-E selection by 3 groups of adult rats offered the following amounts of casein. Group 1 ●——● : period 1 (days 1-14) 0% and 40% casein diets; period 2 (days 15-28), 20% and 60%. Group 2 △——△ : period 1, 0% and 60%; period 2, 10% and 40%. Group 3 ▲——▲ : period 1, 10% and 40%; period 2, 0% and 60% casein diets. The change in diets offered is marked with an arrow.

b) The ratio of the concentration of TRP to the concentrations of the other
   large neutral amino acids (NAA) in plasma can influence brain TRP
   levels : TRP, transported across the blood-brain barrier by the large
   neutral amino acid carrier system, must compete with valine, leucine,
   isoleucine, tyrosine and phenylalanine for entry into the brain
   (Pardridge, 1977). Thus brain TRP concentrations can be predicted by
   plasma TRP/NAA ratios (Fernstrom & Faller, 1978). This is certainly true
   when the TRP/NAA ratio is manipulated over an artificially wide range
   (Fernstrom & Wurtman, 1972), but the effect is much weaker when ratios
   within the normal physiological range are considered (Fig. 5).

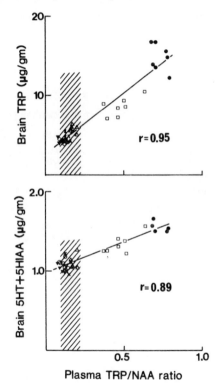

Fig. 5 : A) Correlation between brain tryptophan (TRP) concentration and
the plasma ratio of TRP to the 5 large neutral amino acids (NAA) which
compete with TRP for entry into the brain. B) Correlation between plasma
TRP/NAA and the sum of brain 5-HT and 5-HIAA (5-hydroxyindole acetic acid)
concentrations. Rats were starved for 18 hours then fed either nothing, a
complete amino acid mix or an amino acid mix minus the NAA which compete
with TRP. Controls : 1 hr -o-; 2 hr -▼-. Complete mix : 1 hr -x- 2 hr
-∆-. TRP containing, NAA-free mix : 1 hr -□-; 2 hr -●-. (Fig. adapted
from Fernstrom & Wurtman, 1972). The hatched area on each graph indicates
the range of plasma TRP/NAA ratios in free-feeding rats (see Fig. 8).

c) In rats which have been starved overnight, a high carbohydrate/low
   protein meal increases the TRP/NAA ratio, and it has been suggested that
   this change may also occur in man (Wurtman et al., 1981). These changes
   in plasma TRP/NAA do not occur in free-feeding rats (Ashley et al., 1983)
   nor (Fig. 6) in men eating high carbohydrate morning or evening meals
   (Ashley & Leathwood, 1983). This may well be because the insulin response
   to carbohydrate is more marked after prolonged fasting.

d) Even though a single meal does not influence the plasma TRP/NAA ratio,
   some researchers have found an inverse correlation between % P-E consumed
   during the last 24 hours (or during the last week) and the TRP/NAA ratio
   in plasma (Anderson & Ashley, 1979; Reeves & O'Dell, 1981). Others could
   not repeat these findings (Peter & Harper, 1981).

e) Drug treatments or brain lesions which influence brain 5-HT metabolism
   can change protein choice in the predicted manner, i.e. a sudden rise in
   brain 5-HT preferentially decreases carbohydrate appetite (or increases
   protein appetite). Lowering brain 5-HT has the opposite effect (Arimanana
   et al., 1983; Ashley et al., 1979; Leathwood & Ashley, 1981; Wurtman &
   Wurtman, 1977) although, yet again, inconsistent findings have also been
   reported (Blundell et al., 1979).

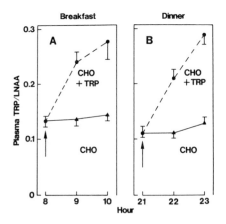

Fig. 6 : Plasma TRP/NAA ratios in men fed, in the morning or evening, a
carbohydrate-rich, protein-free meal ▲——▲; fed the same meal plus
500 Kcal TRP o---o; (from Ashley et al., 1982 and Ashley & Leathwood, 1983).

In summary, this is an attractive hypothesis which might explain how the brain can detect and respond to the level of protein in the diet. Unfortunately, it is based on incomplete and sometimes controversial evidence. As we have shown (see Fig. 4 above) that adult rats can maintain constant levels of protein selection across a change in protein content of the diets offered, they may well be regulating protein intake, so it is interesting to examine to what extent plasma amino acids and brain 5-HT metabolites follow the predictions of the hypothesis.

Table I shows the correlation coefficients between all the relevant food intake, plasma and brain parameters of rats whose protein selection ranged from 15 to 45% P-E (Leathwood & Ashley, 1983). The results confirm the original finding by Ashley & Anderson (1975) that % P-E can predict the plasma TRP/NAA ratio, but show that, when based on individual animals, the correlation is weak and has little predictive value (Fig. 7A).

Protein intake and brain TRP levels were inversely correlated ($p < 0.05$). This correlation remained significant even when any influence of food intake was eliminated by partial correlation analysis. This suggests that protein selection can influence entry of TRP into the brain although, again, the correlation is weak and of poor predictive value (Fig. 7B).

Brain 5-HT, 5-HIAA and 5-HT + 5-HIAA were independent of plasma TRP/NAA, % P-E or protein intake per se (the significant correlations with protein intake disappeared when the influence of food intake was removed by partial correlation analysis). Thus we could not confirm the speculation (Anderson & Ashley, 1979), that because protein selection correlates inversely with plasma TRP/NAA, it also influences brain 5-HT metabolism.

Examination of Figs 5 and 7 show that our results are perfectly coherent with those of Fernstrom & Wurtman (1972). Over the full range, TRP/NAA ratios did predict brain TRP and 5-HT + 5-HIAA (Fig. 5) but within the normal range (Fig. 7) there was no correlation. This suggests that only exceptionally high values of the plasma TRP/NAA ratio can reliably influence brain 5-HT, but within the range produced by normal eating patterns the effect is insignificant.

Table I : Correlation coefficients between food intake, protein intake, carbohydrate intake, % P-E, plasma TRP/NAA, brain TRP, 5-HT, 5-HIAA and brain 5-HT + 5-HIAA

	Food intake	Protein intake	Carbohyd. intake	% P-E	Plasma TRP/NAA	Brain TRP	Brain 5-HT	Brain 5-HIAA
Protein intake	+0.54**	-	-	-	-	-	-	-
Carbohydrate intake	+0.59**	-0.30	-	-	-	-	-	-
% P-E	+0.07	+0.80**	-0.74**	-	-	-	-	-
Plasma TRP/NAA	-0.26	-0.28	+0.14	-0.36*	-	-	-	-
Brain TRP	-0.12	-0.37*	-0.17	-0.31	+0.18	-	-	-
Brain 5-HT	-0.49**	-0.36*	-0.25	-0.12	+0.02	+0.28	-	-
Brain 5-HIAA	-0.53**	-0.38*	-0.29	-0.07	+0.03	+0.65**	+0.70**	-
Brain 5-HT + 5-HIAA	-0.52**	-0.37*	-0.28	-0.19	+0.21	+0.35*	+0.91**	+0.69**

* $p < 0.05$

** $p < 0.01$

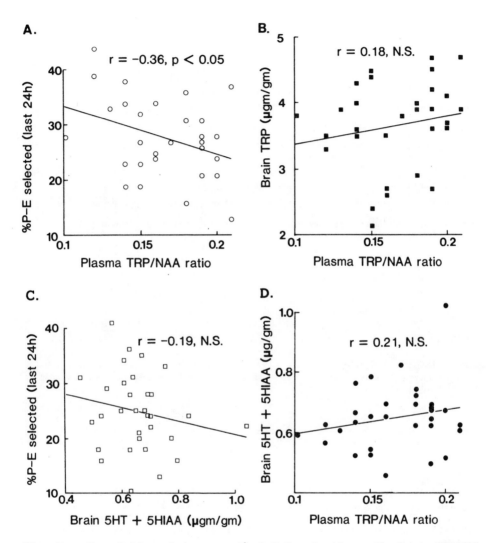

Fig. 7 : Correlations between : A) % P-E selection and plasma TRP/NAA ratio; B) Plasma TRP/NAA and brain TRP; C) Plasma TRP/NAA and brain 5-HT + 5-HIAA; D) % P-E selection and brain 5-HT + 5-HIAA, in adult rats offered a choice of high and low protein diets.

CONCLUSIONS

Animals and men have a variety of mechanisms available to help them maintain nutrient balance. Metabolic adaptation ensures that there is no need for very precise regulation of intake for each nutrient, while food selection may be influenced by internal receptor mechanisms, innate behaviour patterns, learning, social transmission or (in man) cultural practice. According to the circumstances, different combinations or permutations of these may be called into play so that in one experimental situation a particular mechanism may be useful and powerful predictor of behaviour while in others it may be irrelevant.

The first priority of our current research is to define the limits of practical usefulness of the different putative regulatory mechanisms, always bearing in mind that not one of them can have universal application.

REFERENCES

Adolph, W.H. (1944) The protein problem of China. Science 100:1-4.

Anderson, G.H. & Ashley, D.V.M. (1977) Correlation of the plasma tyrosine to phenylalanine ratio with energy intake in self-selecting rats. Life Sci. 21:1227-1234.

Anderson, G.H. & Ashley, D.V.M. (1979) Plasma amino acids, brain mechanisms and the control of protein intake. In : Nutrition in Transition : Proc. Western Hemisphere Nutrition Congress V (Ed. P.L. White & N. Selvey). Am. Med. Assn., Munroe, Wisconsin, pp. 237-246.

Arimanana, L., Ashley, D.V.M., Furniss, D. & Leathwood, P.D. (1983) Protein/ carbohydrate selection in rats following administration of tryptophan, glucose, or a mixture of large neutral amino acids. In : Tryptophan : Proceedings of 4th International Conference on Tryptophan Biochemistry, Pathology and Regulation (Ed. H.B. Schlossberger, W. Kochen, B. Linzen & H. Steinhart) (in press).

Ashley, D.V.M. & Anderson, G.H. (1975) Correlation between the plasma tryptophan to large neutral amino acid ratio and protein intake in the self-selecting rat. J. Nutr. 105:1412-1421.

Ashley, D.V.M., Barclay, D.V., Chauffard, F.A., Moennoz, D. & Leathwood, P.D. (1982) Plasma amino acid responses in humans to evening meals of different nutritional composition. Am. J. Clin. Nutr. 36:143-153.

Ashley, D.V.M., Coscina, D.V. & Anderson, G.H. (1979) Selective decrease in protein intake following drug-induced brain serotonin depletion. Life Sci. 24:973-984.

Ashley, D.V.M. & Leathwood, P.D. (1983) A high carbohydrate, protein-free meal, either at breakfast or in the evening does not alter plasma tryptophan to large neutral amino acid ratios in healthy men. In : Tryptophan : Proceedings of 4th International Conference on Tryptophan Biochemistry, Pathology and Regulation (Ed. H.B. Schlossberger, W. Kochen, B. Linzen & H. Steinhart) (in press).

Ashley, D.V.M., Leathwood, P.D. & Moennoz, D. (1983) A carbohydrate meal increases brain 5-hydroxytryptamine synthesis in the adult rat only after prolonged fasting. In : Tryptophan : Proceedings of 4th International Conference on Tryptophan Biochemistry, Pathology and Regulation (Ed. H.B. Schlossberger, W. Kochen, B. Linzen & H. Steinhart) (in press).

Blundell, J., Latham, C.J., Moniz, E., McArthur, R.A. & Rogers, P.J. (1979) Structural analysis of the actions of amphetamine and fenfluramine on food intake and feeding behaviour in animals and man. Curr. Med. Res. Op. 6:34-54.

Booth, D.A. (1981) The physiology of appetite. Br. Med. Bull. 37:135-140.

Cioffi, L.A., James, W.P.T. & Van Itallie, T.B. (1981) The Body Weight Regulatory System : Normal and Disturbed Mechanisms. Raven Press, New York.

Cohen, M.N. (1977) The Food Crisis in Prehistory. Yale, New Haven.

Das, T.K. & Waterlow, J.C. (1974) The rate of adaptation of urea cycle enzymes, aminotransferases and glutamate dehydrogenase to changes in dietary protein intake. Br. J. Nutr. 32:353-373.

de Garine, I. (1979) Culture et nutrition. Communications 31:70-92.

Denton, D. (1982) The Hunger for Salt. Springer Verlag, Berlin.

Fernstrom, J.D. & Faller, D.V. (1978) Neutral amino acids in the brain : Changes in response to food ingestion. J. Neurochem. 30:1531-1538.

Fernstrom, J.D. & Wurtman, R.J. (1972) Brain serotonin content : Physiological regulation by plasma neutral amino acids. Science, 178:414-416.

Fischler, C. (1981) Food preferences, nutritional wisdom and sociocultural evolution. In : Food, Nutrition and Evolution (Ed. D.N. Walcher & N. Kretchmer). Masson, New York, pp. 59-67.

Galef, B.G. (1976) Social transmission of acquired behaviour : A discussion of tradition and social learning in vertebrates. In : Advances in the study of Behaviour, 6 (Ed. J. Rosenblatt, R.A. Hinde, C. Beer & E. Shaw) Academic Press, New York, pp. 77-105.

Garcia, J., Hankins, W.G. & Rusiniak, K.W. (1974) Behavioural regulation of the "milieu interne" in man and rat. Science 185:824-831.

Harris, L.J., Clay, J., Hargreaves, F. & Ward, A. (1933) Appetite and choice of diet. The ability of vitamin B deficient rats to discriminate between diets containing and lacking the vitamin. Proc. Roy. Soc., London, Ser. B 113:161-190.

Hartsook, E.W., Hershberger, T.V. & Nee, J.C.M. (1973) Effects of dietary protein content and ratio of fat to carbohydrate calories on energy metabolism and body composition of growing rats. J. Nutr. 103:167-168.

Kodicek, E. & Wilson, P.W. (1959) The availability of bound nicotinic acid to the rat. I) The effect of lime - water treatment of maize and subsequent baking into tortilla. Br. J. Nutr. 13:418-430.

Kratz, C.M., Levitsky, D.A. & Lustick, S. (1978) Long-term effects of quinine on food intake and body weight in the rat. Physiol. Behav. 21:321-324.

Krebs, H.A. (1964) The metabolic fate of amino acids. In : Mammalian Protein Metabolism (Ed. H.N. Munro & J.B. Allison). Academic Press, New York, pp. 125-177.

Lat, J. (1967) Self-selection of dietary components. In : Handbook of Physiology, Sect. 6 - Alimentary Canal. Vol. 1 : Control of Food and Water Intake (Ed. C.F. Code & W. Heidel). Am. Physiol. Soc., Washington D.C., pp. 367-386.

Leathwood, P.D. (1978) Influence of early undernutrition on behavioural development and learning in rodents. In : Studies on the Development of Behaviour and the Nervous System. Vol. 4 : Early Influences (Ed. G. Gottlied). Academic Press, New York, pp. 187-209.

Leathwood, P.D., Arimanana, L., Resenterra, P. & Ashley, D.V.M. (1981) Defense of energy intake and protein intake in the rat. Poster presented at the XIIth International Congress of Nutrition. San Diego, California.

Leathwood, P.D. & Ashley, D.V.M. (1981) Nutrients as regulators of food choice. In : The Body Weight Regulatory System - Normal and Disturbed Mechanisms (Ed. L.A. Cioffi, W.P.T. James & T.B. Van Itallie). Raven Press, New York, pp. 263-269.

Leathwood, P.D. & Ashley, D.V.M. (1983) Strategies of protein selection in weanling and adult rats. Appetite (in press).

Le Magnen, J. (1971) Regulation and control of food intake. In : Progress in Physiological Psychology, 4 (Ed. E. Stellar & J.M. Spragues). Academic Press, New York, pp. 204-259.

Leprohon, C.E. & Anderson, G.H. (1982) Relationships among maternal diet, serotonin metabolism at weaning and protein selection of progeny. J. Nutr. 112:29-38.

Leshner, A.I., Seigel, H.I. & Collier, G. (1972) Dietary self-selection by pregnant and lactating rats. Physiol. Behav. 8:151-154.

Munro, H.N. (1964) General aspects of the regulation of protein metabolism by diet and hormones. In : Mammalian Protein Metabolism (Ed. H.N. Munro & J.B. Allison). Academic Press, New York, pp. 125-177.

Musten, B., Peace, D. & Anderson, G.H. (1974) Food intake regulation in the weanling rat : Self-selection of protein and energy. J. Nutr. 104:563-572.

Oddoye, E.A. & Margen, S. (1979) Nitrogen balance studies in humans : Long-term effect of high nitrogen intake on nitrogen accretion. J. Nutr. 109:363-377.

Pardridge, W.M. (1977) Regulation of amino acid availability to the brain. In : Nutrition and the Brain (Ed. R.J. Wurtman & J.J. Wurtman). Raven Press, New York, pp. 141-204.

Peters, J.C. & Harper, A.E. (1981) Protein and energy consumption, plasma amino acid ratios and brain neurotransmitter concentrations. Physiol. Behav. 27:287-298.

Reeves, P.G. & O'Dell, B.L. (1981) Short-term zinc deficiency in the rat and self-selection of dietary protein. J. Nutr. 111:375-383.

Richter, C.P. (1936) Increased salt appetite in adrenalectomised rats. Am. J. Physiol. 115:155-161.

Richter, C.O. & Barelare, B. (1938) Nutritional requirements of pregnant and lactating rats studied by the self-selection method. Endocrinology 23:15-24.

Rodgers, W. & Rozin, P. (1966) Novel food preferences in thiamine-deficient rats. J. Comp. Physiol. Psychol. 61:1-4.

Rolls, B.J. & Rolls, E.T. (1981) The control of drinking. Br. Med. J. 37:127-130.

Rozin, P. (1967) Thiamine specific hunger. In : Handbook of Physiology, sect. 6 - Alimentary Canal. Vol. 1. Control of Food and Water Intake (Ed. C.F. Code & W. Heidel). Am. Physiol. Soc. Washington D.C., pp. 411-431.

Rozin, P. (1968) Are carbohydrate and protein intakes separately regulated ? J. Comp. Physiol. Psychol., 65:23-29.

Rozin, P. (1976) The selection of foods by rats, humans and other animals. In : Advances in the Study of Behaviour, 6 (Ed. J. Rosenblatt, R.A. Hinde, C. Beer & E. Shaw). Academic Press, New York, pp. 21-76.

Rozin, P. & Fallon A.E. (1981) The acquisition of likes and dislikes for foods. In : Criteria of Food Acceptance (Ed. J. Solms & R.L. Hall). Forster Verlag, Zurich, pp. 35-48.

Rzoska, J. (1953) Bait shyness, a study in rat behaviour. Br. J. Anim. Behav. 1:128-135.

Scott, E.M. & Vernay, E.L. (1947) Self-selection of diet VI : The nature of appetites for the B vitamins. J. Nutr. 34:471-480.

Solms, J. and Hall, R.L. (Eds) (1981) Criteria of food acceptance : How man chooses what he eats. Forster, Zurich, Switzerland.

Steiner, J. (1973) The human gustofacial response. In : Fourth Symposium on Oral Sensation and Perception - Development of the Fetus and Infant (Ed. J.F. Bosma). U.S. Government. Print Office, Washington D.C., pp. 254-278.

Woodger, T.L., Sirek, A. & Anderson, G.H. (1979) Diabetes, dietary tryptophan and protein intake regulation in weanling rats. Am. J. Physiol. 236:R307-R311.

Wurtman, R.J., Hefti, F. & Melamed, E. (1981) Precursor control of neurotransmitter synthesis. Pharmacol. Rev. 32:315-335.

Wurtman, J.J. & Wurtman, R.J. (1977) Fenfluramine and fluoxetine spare protein consumption while suppressing caloric intake by rats. Science 198: 1178-1180.

Wurtman, J.J. & Wurtman, R.J. (1979) Drugs that enhance central serotoninergic transmission diminish elective carbohydrate consumption by rats. Life Sci. 24:895-904.

Wyrwicka, W. (1976) The problem of motivation in feeding behaviour. In : Hunger - Basic Mechanisms and Clinical Implications (Ed. D. Novin, W. Wyrwicka & E. Bray). Raven Press, New York, pp. 203-213.

THE RELATIONSHIP OF PELLAGRA TO CORN AND THE LOW AVAILABILITY OF NIACIN IN CEREALS

Kenneth J. CARPENTER

University of California, Department of Nutritional Sciences,
Berkeley, CA 94720, U.S.A.

SUMMARY

The poorest inhabitants of an area generally eat the narrowest range of foods, and one staple (which serves as a cheap source of calories) dominates. In turn, the specific type of malnutrition seen in that area depends upon that predominant staple and how it is processed before consumption.

Corn, used here in the sense of "Indian corn" or maize, was brought to Europe from America, and over the period 1750-1850 became the typical peasant's staple in many of the areas bordering the Mediterranean. By the end of that period, it had also come to be recognized that pellagra had become a serious, chronic disease in these same countries, flaring up each spring amongst the poorest people living on diets containing much corn and very little animal food (i.e., meat, eggs or dairy products) or wine and being generally in a state of wretchedness. Nothing of the sort was seen in areas where wheat and rice were the staple foods, even when they were highly milled.

Most scientists agreed on this association with corn, though not on what was the true cause-and-effect relationship. Research

in the present century has shown that pellagra is primarily due
to a dietary deficiency of niacin. However, the niacin content
of different foods did not tie in well with their pellagra-
preventive value. But then it was discovered that a second
nutrient, tryptophan, could act as precursor of the vitamin with
approximately one sixtieth of the activity of the actual
vitamin. The "niacin equivalent" values of foods (calculated
from their content of both nutrients) show a much better correlation with their pellagra-preventive value. Thus, mature corn
is lower in niacin content than are wheat and rice; also the
mixed proteins of corn are lower in their tryptophan content.

What is not explained by the calculation of "niacin equivalent" is the general freedom from pellagra of the peasants in
Mexico and Central America, where corn has been the staple for
millenia and where poverty, the consequent lack of animal foods
in the diet, and general misery, have been fully equal to the
conditions in Europe.

It has been known for 40 years that analytical values for the
niacin content of foods depended greatly on the method of
extraction used, with the highest values being obtained after
treatment with alkali. We have confirmed with rat growth assays
that the niacin in corn, wheat and rice is only about one-third
available to this species, even after ordinary cooking at
neutral pH. Cooking corn to make the traditional "tortillas"
with calcium hydroxide resulted in all the niacin becoming
available. Though some was leached out at the soaking stage, the
net niacin value of the corn was still significantly improved.

A concentrate of "bound niacin" prepared from wheat bran,
using a procedure developed by Kodicek and Mason at Cambridge,
was estimated to be about 25% available to humans. The assay was
based on the level of extra urinary metabolites, N^1-methyl-
nicotinamide and N^1-methyl-2-pyridone-5-carboxamide after test
doses of the preparation. The same preparation treated with

calcium hydroxide before dosing was apparently about 70% available as compared with a standard of nicotinic acid.

We are planning more human studies with corn products cooked
in different ways. It appears that the traditional processing
may have been a crucial factor in making corn a safe staple,
while its significance went unrecognized by Europeans.

* * *

The poorest inhabitants of an area where the standard of living is low
generally eat the narrowest range of foods, and one staple (which serves as
the cheapest local source of calories) dominates. In turn, the specific type
of malnutrition seen in that area depends upon the particular predominant
staple and how it is processed before consumption. The use of white rice has
been repeatedly associated with beriberi, as has corn with pellagra.

Pellagra in Europe

Corn, used here in the sense of "Indian corn" or maize, was brought to
Europe from America, and over the period 1750-1850, became the typical
peasant's staple in many of the areas, bordering the Mediterranean and also
in parts of Rumania and Southern Russia. By the end of that period, it had
also come to be recognized that pellagra had become a serious, chronic
disease in these same areas, flaring up each spring amongst the poorest
peasants living on diets containing much corn and little animal food (i.e.,
meat, eggs or dairy products) or wine and being generally in a state of
wretchedness. The characteristic signs of the disease were severe dermatitis
of those parts of the skin exposed to the sun, leading to dry, fissured
scabs, also chronic diarrhoea and a mental deterioration marked by
withdrawal and depression (e.g. Thiery, 1932; Aykroyd et al., 1935). Nothing
of the sort was seen in areas where wheat and rice were the staple foods,
even where they were highly milled.

In the second half of the nineteenth century, the majority of investigators probably agreed with Roussel (1845) who summed up as follows : 1) The

immediate cause of the disease is damaged corn. We must, therefore, take measures to ensure that only healthy corn of good quality enters into consumption. 2) The principal predisposing cause is an inadequate diet which is almost exclusively vegetable. The proportion of animal products in the diet of the peasants must be considerably increased.

H.F. Harris (1919), a Georgia physician, wrote a scholarly book reviewing the experiments and conclusions of European investigations into pellagra up to that date. He personally became convinced by the evidence that the disease was caused by a toxin in mouldy corn. Schoental (1980) is still of this opinion. One of the features of pellagra was that it flared up mostly in spring. This was explained on the "mould" theory by the corn meal's having become increasingly mouldy during its storage over the months since harvest. The restriction of the disease to poorer people was explained by the rich being able to discard corn meal which was no longer in "fresh" condition because they could afford to purchase more.

Pellagra in the U.S.A.

In 1906, pellagra began to be a serious problem also in the South-Eastern portion of the U.S.A. Searcy (1907) who studied the first outbreak of 88 cases in an insane asylum in Alabama, concluded that it came from the use of mouldy corn bread. After wheat and potatoes had replaced corn bread and grits in the diet, no new cases occurred. When the disease began to be found in one state after another with an epidemic-like appearance, a second theory gained ground, namely that it was an infection, probably transmitted by insects eating contaminated excreta (Sambon, 1910). Epidemiological studies confirmed that where it occurred in a town, the incidence was higher amongst people who had primitive and insanitary privies (Siler et al., 1914). Although corn was a dietary item in all the pellagra areas, investigation turned up no evidence of unusual mould infestation on the corn being eaten by the affected groups (Alsberg, 1909). Nor did dietary surveys show a higher consumption of corn by pellagrins than by those who remained healthy (Goldberger et al., 1918b). In all the diets studied, consumption of corn products was less than that of wheat products (Goldberger et al., 1918b; Stiebeling & Munsell, 1932).

Because of those observations, corn ceased to be regarded in the U.S.A. as having any specific relation to pellagra. Goldberger and his U.S. Public Health Service (USPHS) team discovered one consistent difference in the diet of pellagrins - they had very little milk and cheese compared with those who remained healthy (Goldberger et al., 1918b). The preventive and curative value of milk was confirmed in trials in orphanages and mental hospitals (Goldberger et al., 1915; Goldberger & Tanner, 1922).

Vitamin deficiency

At first the value of milk was thought, by the USPHS group, to lie in its amino acid content since neither its mineral fraction nor cod-liver oil and orange juice (the materials regarded as "vitamin-rich" at that time) were found to be active (Goldberger & Tanner, 1922). Then the high potency of yeast extracts was discovered (Goldberger & Tanner, 1925) and their opinion changed again to pellagra's being a B-vitamin deficiency.

Dogs proved useful model animals; a pellagrous human diet induced a disease-condition, blacktongue, with some striking similarities to pellagra (Goldberger & Wheeler, 1928). They had, in fact, provided Goldberger with the first evidence of the potency of yeast extracts (Goldberger et al., 1918a). After Goldberger's death in 1929, and when nicotinic acid had been shown to be both a constituent of a co-enzyme (Warburg & Christian, 1935) and a growth factor for bacteria (Knight, 1937), workers at Wisconsin University tested nicotinic acid and found that it cured dogs with black-tongue (Elvehjem et al., 1937). It also gave dramatic responses with cases of pellagra (Spies et al., 1939). Some of the symptoms were not cleared up in a proportion of cases and these responded to riboflavin and/or thiamin.

In Goldberger's basal diets for both his human and dog studies, corn formed a much larger proportion (more than 50% of total dry matter) than had been recorded in dietary surveys in the U.S. This was not considered to have any particular significance at the time. However, after the discovery of the vitamin activity of nicotinic acid, the Wisconsin group attempted to produce a corresponding deficiency condition in the rat. Surprisingly, rats showed no requirement for nicotinic acid on the usual purified diet with casein as the protein source. They continued to grow and even to excrete the vitamin in their urine. But when this diet was diluted with 40% corn meal, the rats did become deficient (Krehl et al., 1945). No such effect was seen when wheat or rice was the diluent (Krehl et al., 1946). Here apparently was a dramatic return to the "zeism" theory of pellagra with corn having a specific malevolent influence.

One difference between corn and the other cereals is that it has a lower content of tryptophan (expressed as g/100 g protein) and it was found that the depressed growth of the rats on the "40% corn" diet was corrected by supplementary tryptophan as well as by nicotinic acid. Similarly a growth depression could be obtained without corn by adding gelatin (a protein lacking tryptophan) or even by adding an amino acid mix which omitted tryptophan (Krehl et al., 1945). The activity of tryptophan also provided an explanation for the pellagra-preventive value of animal protein foods of low niacin content.

Because of the great difference between the indole ring in the amino acid and the pyridine ring in the vitamin, it was thought at first that tryptophan must stimulate production of nicotinic acid indirectly, probably by way of the intestinal microflora. Then it was shown by experiments with isotopes that nicotinic acid was being formed from tryptophan (Heildelberger et al., 1948). The complicated succession of enzymes required for this conversion was worked out over the next 15 years (Nishizuka & Hayaishi, 1963).

Experiments with humans have shown that about one molecule in 30 of supplementary tryptophan is converted to "niacin" (i.e. the general term for the vitamers nicotinic acid and nicotinamide) (Goldsmith et al., 1961). Because of the differences in molecular weight, this is equivalent to approximately 1 mg niacin per 60 mg tryptophan. For calculating the "nicotinic acid equivalents" (NE) in a diet, it is now conventional to add 1/60th of the weight of tryptophan present to the weight of performed niacin (National Research Council, 1980). This proportion only represents an average conversion factor and it has been shown to be affected by hormonal influences (e.g. Knox & Auerbach, 1955). There are also important species differences.

Species differences

It is now possible to understand why animal models were at first so misleading to investigators. McCollum had been the pioneer, demonstrating the usefulness of rats as models for producing vitamin A deficiency on diets similar to those with which the disease condition had been observed in calves (McCollum & Davis, 1915). When others (e.g. Funk, 1912) suggested that scurvy and pellagra might also be vitamin deficiency diseases, McCollum's research group fed rats on diets similar to those associated with each disease, but in neither case did the rats fail to thrive. He concluded that scurvy was, in reality, a disease of autointoxication (McCollum & Pitz, 1917) and that pellagra was an infectious disease analogous to tuberculosis (McCollum et al., 1919).

We know now, of course, that the rat needs vitamin C for its metabolism but that, unlike humans, it can synthesize it from glucuronic acid (Review by Sato & Udenfriend, 1978). The rat also needs less tryptophan in its diet (as a percentage of total calories) to obtain enough niacin than do humans. Even 10% casein in a diet usually supplies enough tryptophan for a reasonable synthesis of niacin. Also, the deficient rat does not show dermatitis or diarrhoea.

The dog, on the other hand, which Goldberger was able to use as his animal model, needs more tryptophan if it is to be independent of a supply of dietary niacin. It can, for example, receive as much as 18% of its calories in the form of casein and still become niacin-deficient (Singal et al., 1948). At the extreme is the cat family which can apparently make little or no use of tryptophan as a precursor of niacin (Carvalho Da Silva et al., 1952).

Bound niacin in cereals

A second confusing factor in trying to evaluate a diet for its "niacin value" (or "niacin equivalent" value) is the uncertainty concerning the bioavailability of the niacin in different foods.

Soon after the discovery of the "vitamin" nature of nicotinic acid, it was realized that microbiological assays of the nicotinic acid value of cereal foods were giving different results according to the methods of extraction used (Kodicek, 1940; Krehl & Strong, 1944). Further studies confirmed that a considerable portion of the niacin in a range of cereals was in a "bound" or "precursor" form, and was released to become available to the assay micro-organism by refluxing in either acid or alkali (Kodicek & Pepper, 1948; Ghosh et al., 1963).

The relative quantities of free and bound niacin were first measured chemically in a semi-quantitative procedure, running paper chromatograms of extracts of food, and then developing the niacin-containing spots with cyanogen bromide and 4-aminobenzoic acid (Kodicek & Wilson, 1959). Thin-layer chromatograms on plates coated with silica gel gave more reproducible Rf values (Carter & Carpenter, 1981). The procedure was still only semi-quantitative but served very well to differentiate foods with most of their niacin in bound form from those with it almost entirely free. A quantitative separation procedure in which both free and bound could be determined was developed using Sephadex G-10 (Carter & Carpenter, 1981). The successive fractions eluted from the column were digested in alkali and their free nicotinic acid (NA) content determined. The results of a typical run for jowar (an Indian variety of Sorghum vulgare) are shown in Fig. 1.

It was found, in addition, that alkaline treatment with NaOH increased the vitamin value of such extracts for rats (Chaudhuri & Kodicek, 1960). It was also reported that replacing raw corn meal in a niacin-deficient rat diet with Mexican-style, lime-cooked corn resulted in greatly improved growth, nearly equivalent to that obtained from supplementing the raw corn diet with an optimal level of niacin (Laguna & Carpenter, 1951; Kodicek & Wilson, 1959). Others confirmed this finding but suggested that the effect was to be explained by a change in the digestibility of the corn proteins and its effective amino acid balance rather than by the nicotinic acid being brought into a more available form (Massieu et al., 1956; Squibb et al., 1959). This has been the subject of a number of studies and it now seems agreed that the effect is, at least largely, due to the conversion of bound niacin to more available forms (Harper et al., 1958; Kodicek & Wilson, 1959; Christianson et al., 1968).

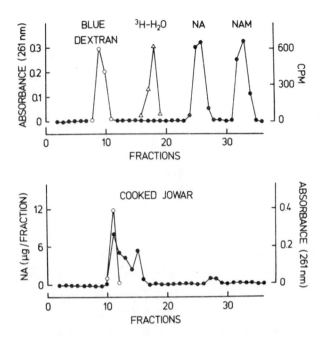

Fig. 1 : Elution patterns when standard mixtures or sorghum extracts were eluted from Sephadex G-10 columns. Absorbances for Blue Dextran (o--o) and NA (●--●) and cpm for $^3H-H_2O$ (Δ--Δ).

It was reported in one paper that the same improvement could be obtained by cooking corn at neutral pH (Pearson et al., 1957). This has not been confirmed in a series of other trials (Harper et al., 1958; Carpenter et al., 1960). The original report may have been related to the final drying being done under such drastic conditions that there was considerable destruction of lysine. This deserves re-examination.

These questions are, of course, of great interest in relation to the paradox that pellagra has been associated with corn consumption amongst the poorest portion of the population in every area except Mexico and Central America, where it has been the staple food for millenia (Salas, 1863; May & McLellan, 1972). Is this freedom to be explained by the traditional cooking of corn at an alkaline pH with lime-water to make tortillas, or are other explanations, particularly the use of beans as a supplement to corn, the real reasons ? We will return to this question later.

Several laboratories have attempted to isolate the bound form of niacin from cereals so as to study its chemical characteristics (Das & Guha, 1960; Kodicek & Wilson, 1959; Christianson et al., 1968; Mason et al., 1973; Mason & Kodicek, 1973). No one has reported obtaining a single molecule of identified structure. All are agreed that the nicotinic acid is part of either a complex structural carbohydrate or a glycoprotein. It may be that a family of molecules is involved. The indication from the work of Mason & Kodicek (1973) is that the nicotinic acid is esterified to a glucose unit which is, in turn, attached to other units. Recently, Sandhu & Fraser (1981) have shown that nicotinoyl cellulose has properties similar to those of naturally occurring "bound niacin" when fed to deficient rats.

Biological assays

Concentrates of bound niacin, purified by successive fractionation may not be representative of all the "bound" niacin in a food. The very procedures of fractionation may also alter its properties. These were two reasons for our wishing to carry out quantitative rat bioassays for the niacin value of whole foods (Carter & Carpenter, 1982a). The main problem is to keep the balance of tryptophan and other amino acids in the test diets constant so that it does not influence the response of the rat when test foods are added to a niacin-deficient diet.

We used weight gain or "gain/g food eaten" of weanling rats as our response measure. This required careful choice of tryptophan level for the diets so that there was enough to permit the protein synthesis required for rapid growth when a high level of niacin was added, but not so much that it supplied enough of the vitamin, in the absence of added niacin, to support more than very slow growth. The composition of the most satisfactory diet that we arrived at is shown in Table I.

Our final procedure has been to feed weanling (i.e. 21-day-old) male rats for 4 days on the basal diet and then randomize them to the test diets for an experimental period of 18-20 days. On the basal diet without a supplement, their weight remained virtually constant; with a supplement of 12 mg/kg nicotinic acid, they gained at a rate of approximately 4 g per day.

Table I : Basal diet for niacin rat assays (g/kg)

Casein	70
Gelatin	65.5
Amino acids	varied*
Vitamins (-Niacin)	5
Minerals	55
Corn oil	30
Sucrose	ad 1,000

* Tryptophan made up to 0,97 g/kg diet. Other A.A.S.
  to constant values meeting full requirements.

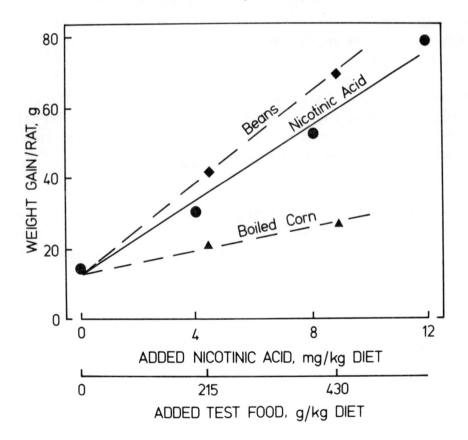

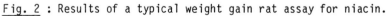

Fig. 2 : Results of a typical weight gain rat assay for niacin.

In order to obtain data that could be subjected to a valid "slope-ratio" statistical analysis, we used two levels of each test material and three levels of standard in every assay (Finney, 1964). Every test diet was fed to four cages each containing two rats. The calculated coefficients of variation of the estimates varied according to the degree to which the response to the test food covered the range of straight-line response to the standards. The higher values were obtained with samples where the response of the rats was very small. An example is shown in Fig. 2, where the c.v. of the estimate for beans was 9 and for boiled corn was 26%. One would like to have lower values for error, but these are rather typical of results from bioassay.

Results for 10 foods are summarized in Table II. For none of the test items that had the bulk of their niacin in the free from was the rat assay value significantly different from the niacin content determined chemically. Wheat, milo and corn had only trace amounts of free niacin, but rice and potatoes apparently had about one-quarter of their niacin free. All these samples gave rat assay values well below the values for total niacin content. On the hypothesis that all the free niacin was available, we estimated the availability of the bound portion from how much the rat value exceeded the chemical value for free niacin. As we see in Table II, the estimates for the four cereals averaged 32%.

From experiments where foods rich in bound niacin have been added to niacin-deficient diets without adjustment of niacin balance (e.g. Chaudhuri & Kodicek, 1960; Christianson et al., 1968), the general finding has been that the bound niacin is of no value to young rats. We are not sure why our results differ. One possibility is that some other characteristic of the test foods is stimulating growth slightly in the young rats. Mason & Kodicek (1970), working with a concentrate of soluble "bound niacin" from wheat bran, with the free niacin removed by dialysis, found that it too gave a growth response with rats. We made a similar preparation and a rat assay gave a value equivalent to an estimated 17% availability of the niacin (Carter & Carpenter, 1982a).

Table II : Rat assay results compared with chemical values
for free and total niacin

| | Niacin values (as nicotinic acid), mg/kg | | |
	Chemically free	Total	Rat assay
Cereal foods :			
Boiled milo	ND*	45.5	16 ± 3.8
Boiled wheat	ND	57.3	18 ± 5.4
Concentrate from wheat bran	ND	2,800	480 ± 240
Boiled rice	17	70.7	29 ± 6.6
Boiled corn	1.1	18.8	6.8 ± 1.8
Tortillas	11.7	12.6	14 ± 1.8
Steamed sweet corn	45	56.4	48 ± 7.2
Non-cereal foods :			
Baked potatoes	12	51	32 ± 6.9
Baked liver	297	306	321 ± 30
Freeze-dried instant coffee	315	597	417 ± 67
Baked beans	19	24	28 ± 2.7

* ND = not detected

Niacin assay with human subjects

We went on to assay another batch of this material with human subjects, to obtain an indication as to whether humans behaved like rats in their metabolism of bound niacin (Carter & Carpenter, 1982b). The advantage of using a concentrate for this purpose was that we did not need to make adjustments for significant contributions of tryptophan or other amino acids.

There could, of course, be no question of making human subjects deficient. We, therefore, followed the common procedure of estimating vitamin status from urinary excretion of metabolites (Melnick et al., 1945). Human assays are also expensive but we hoped to assay 3 test materials on each of 3 subjects within a total test period of 40 days.

As with the rat assay, the basal diet had to provide a constant daily intake of available niacin and also of tryptophan and the other essential amino acids. We chose a level of 20 mg niacin equivalent intake per day. This met the U.S. "Recommended Dietary Allowance" and was also expected to correspond to a point on the dose-response curve for urinary metabolites, where, with an additional intake of available niacin, a large proportion would be recovered in the urine. The diets were chosen to be as palatable as possible but to omit items that could be a serious source of variation of "niacin-equivalent" intake. Thus, meat and fish were omitted and bread was made for us from non-enriched white flour. The midday meal consisted of a "formula" drink that served as a vehicle for providing supplementary minerals and vitamins other than niacin.

We wanted our test doses to have niacin contents such that the total daily intake was still within the normal range. Results with very large supplements indicating a low percentage availability could be criticized for being "non-physiological". Also, in order to have the test as relevant as possible to practical conditions, one portion of the concentrate was mixed with corn starch and water and baked at neutral pH, as for most cereal foods, and another portion was given the same treatment but with added calcium hydroxide (as in the production of Mexican and Central American tortillas). The second treatment resulted in the niacin being released from its bound form(s) and served as one positive control. The third material tested was pure nicotinic acid. The subjects were all scientists continuing their normal activities, but experienced in quantitative measurement and with a great interest in the work.

From a scrutiny of the literature and from a short preliminary test with one subject, it appeared that a 14-day preliminary period on the basal diet should bring urinary excretion down to a plateau level, and 24-hour urines were collected form the 6th day on. On days 14 and 15, each subject received one of the test doses spread over the two days, each total dose containing

35 mg niacin (except for the pure nicotinic acid which was only 24 mg).
Urine continued to be collected each day. It was expected (and the results
confirmed it) that the urinary response to the doses would be complete in
4-5 days and that excretion would return to plateau levels after that. The
subjects were then given their second dose on days 22-23 and their third
dose on days 30-31. The doses were given in a different order to each
subject.

Each 24-hour sample of urine, was analyzed in duplicate for the two main
niacin metabolites, N^1-methylnicotinamide and N^1-methyl-2-pyridone-5-
carboxamide and the total of the two calculated as niacin equivalents. A
typical set of results for one subject is shown in Fig. 3. The above-plateau
response summed for the six days following each dose is set out in
Table III. We see that the differences between treatments were significant.
When the urines were analyzed for their creatinine content and the daily
excretion of metabolites expressed "per mg creatinine" the residual variability was reduced, and the treatment differences became highly significant.

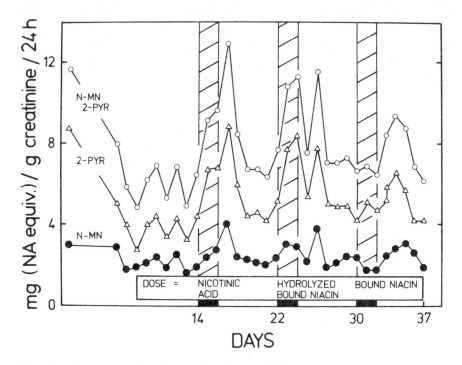

Fig. 3 : Response of urinary niacin metabolites to test doses
containing niacin.

Table III : Baseline values for excretion of N-MN + 2-PYR
and the net additional excretion after each dose*

Subject	Mean baseline excretion	Dose (total NA content, mg)		
		NA (24.0)	ABN (35.3)	BN (35.3)
	mg/24 h	Additional excretion of NA equivalents, mg		
1	(12.8)	24.1	20.3	7.38
2	(8.24)	17.1	25.7	11.30
3	(3.86)	22.6	19.6	6.30
Means*		21.3[a]	21.9[a]	8.33[b]
		Daily mg NA equivalents/g creatinine		
Means*		15.9[a]	16.6[a]	6.34[b]

* NA, nicotinic acid; BN, bound nicotinic acid; ABN, alkali-treated BN.
  Values with different superscripts are significantly different.

Since the response to the bound niacin preparation was equivalent to only 24% of the dose and the average response to the two doses containing only free niacin was 76%, it was concluded that the phenomenon of "bound niacin" in a variety of foods, being of low availability to rats could be duplicated in humans. To go beyond this and estimate the availability of the bound niacin as "100 x 24/76 = 32%", would assume linear responses to both standard and test material. We can only say, for certain, that the availability in the test material is lower to a highly significant extent. An experiment with one test material does not, of course, provide proof that all the assay values obtained with rats are applicable to humans, but it does suggest that, until further assays with humans are conducted, it would be prudent to use the lower rat assay values when calculating the adequacy of the niacin supply in human diets rather than the total niacin values from chemical analyses that are used at present.

Re-evaluation of pellagrous diets

Table IV sets out representative "niacin-equivalent" values for a number of foods and shows the effect of allowing for an estimated 35% availability of the bound niacin. Whether one considers total NE or available NE, it is clear that corn (maize) has a considerably lower value than the other cereals. It is also clear that the pellagra-preventive (PP) value of dairy products is explained by their contribution of tryptophan, even though they are quite low in niacin.

Table IV : The total and available "niacin-equivalent" (NE) values (mg/10^3 kcal) of common foods

Foods		Total NE	Available NE[b]
Potatoes		23 (0.23)[a]	17 (0.32)
Whole cereals	Wheat	21 (0.39)	12.5 (0.65)
	Rice	24 (0.23)	11.5 (0.47)
	Sorghum	15 (0.33)	8.5 (0.59)
	Corn	10 (0.32)	5.3 (0.59)
Refined cereals	Wheat	7.5 (0.66)	5.7 (0.85)
	Rice	9.0 (0.34)	4.9 (0.61)
	Corn	5.5 (0.35)	3.1 (0.62)
	Tortillas	6.2 (0.41)	
	Beans	19 (0.67)	
	Meat	26 (0.51)	
	Milk	15 (0.92)	
	Eggs	22 (0.94)	

(a) The proportion of the NE coming from tryptophan is shown in parentheses.
(b) The niacin in the cereals (except tortillas) is assumed to be only 33% available and in the potatoes 63%.

If it is the case that most of the niacin in cereals (other than those cooked in alkali) is unavailable for humans, does this provide an explanation for which diets are and which are not pellagragenic ? Certainly it helps. Table V summarizes newly calculated estimates of the composition of diets from old surveys (Lewin, 1980; Carpenter, 1981a).

Table V : Composition and calculated nutritional value of published average diets for different communities (from Carpenter, 1981a)

	Rumanian peasant, 1935 (pellagrous)	Guatemalan rural, 1965 (healthy)	Carolina mill villages, 1916 (healthy)	(pellagrous)
Percentage of calories from different foods :				
Corn	75.5	57.3	15.7	16.1
Other cereals	9.9	7.7	38.6	39.5
Beans	8.6	8.5	2.8	3.9
Dairy products	0.3	3.3	7.2	1.9
Meat and eggs	2.7	4.8	4.8	5.1
Fruit and vegetables	3.0	3.8	3.3	3.5
Fats	-	3.6	22.3	23.4
Sugar and syrups	-	10.0	5.3	6.6
Calculated nutrient value of diets, expressed per thousand kcal :				
Protein, g	30.8	29.2	27.4	25.7
Tryptophan, mg	271	280	340	278
Niacin, mg[1]	7.3 (3.4)	6.5 (6.3)	4.3 (3.4)	4.1 (3.0)
Niacin equivalents, mg[1]	11.8 (7.9)	11.2 (11.0)	10.0 (9.0)	8.7 (7.6)

(1) Niacin-equivalents are calculated as "Tryptophan/60" plus niacin. Values in brackets are calculated by assigning the whole cereals used in Rumania an availability of 32%, the refined cereals used in Carolina 40% and the lime-cooked corn used in Guatemala 100%. The Guatemalan niacin value has 1.0 mg/day added as the contribution from coffee.

The typical Rumanian peasant diet, when pellagra was a serious problem, was estimated to contain 11.8 mg total niacin equivalents (NE) per 1,000 kcal. This was more than the corresponding value of 11.2 mg for a Guatemalan rural diet where the disease did not occur. Adjusting for the assumed low availability of bound niacin gives estimates of 7.9 and 11.0 mg respectively of available NE per 1,000 kcal. The Rumanian value of 7.9 is similar to a value of 7.6 mg available niacin calculated for pellagrous family diets in the U.S.A. in the original USPHS survey (Goldberger et al., 1918b).

The odd thing is that work carried out since 1945 to determine human NE requirements has given results (Goldsmith et al., 1955; Horwitt, 1958) from which it was concluded that diets containing 4.4 mg total NE per 1,000 kcal were just adequate and that 6.6 mg was a safe recommended dietary allowance (RDA) (National Research Council, 1980).

This is a paradoxical situation. From a re-examination of Goldberger's later studies of subjects receiving experimental diets, calculations again indicate that many of those developing pellagra were apparently receiving as much as 7.9 mg available NE per 1,000 kcal (W.J. Lewin & K.J. Carpenter, unpublished study). On the other hand, these diets and the family diets of pellagrins surveyed in the U.S.A. (Goldberger et al., 1918b; Stiebeling & Munsell, 1932) are all apparently well below the RDA of 0.6 mg/1,000 kcal for riboflavin.

Sebrell & Butler (1938) showed that some of the deficiency signs on Goldberger's "pellagra-producing" diets were cleared up with supplementary riboflavin, and probably the critics of Goldberger's first production of "experimental pellagra" were right in thinking that a different condition had been produced which we would now attribute to deficiency of riboflavin (cf. Carpenter, 1981b).

However, that it not the whole story, for the use of nicotinic acid did have a dramatic effect in the treatment and prevention of pellagra in the U.S.A. Could it be that marginal deficiencies of other B-vitamins were increasing the requirement for niacin, perhaps through a reduced efficiency of the tryptophan-to-niacin pathway ? The complex interrelationships of the B-vitamins have been the subject of many papers. The "pellagra-preventive" food supplements used by the USPHS in their classic studies were all sources of many nutrients. We are in the middle of testing different hypotheses about vitamin interactions to account for their relative activities (W.J. Lewin & K.J. Carpenter, work in progress).

The cases of pellagra studied at the National Institute of Nutrition (NIN), Hyderabad, have been of people, mostly men, whose staple cereal food was sorghum, with the local variety name of "jowar". Sorghum has its niacin in bound form, as already discussed, but its protein is of relatively high tryptophan content, i.e. like wheat and rice rather than corn. The obvious characteristic that sorghum shares with corn is a high leucine content. In view of this, Gopalan and Srikantia (1960) hypothesized that it was the high leucine content of corn rather than its low tryptophan that was mainly responsible for its association with pellagra. Since then, there has been a series of papers from NIN indicating the pellagragenic activity of leucine both in human subjects and in monkeys and dogs (cf. review by Gopalan & Jaya Rao, 1975).

It has been a continuing paradox that no such effect of leucine has been found in studies reported from other countries (e.g. Manson & Carpenter, 1978; Nakagawa & Sasaki, 1977; Patterson et al., 1980; Cook & Carpenter, 1982). No study has been reported of the communities to which the Indian pellagrins belong - i.e. incidence by sex, age and season and the composition of family dietaries. The explanation for the Indian results may lie in the finding of vitamin B_6-deficiency in the subjects and/or a connection with alcoholism (Bapurao & Krishnaswamy, 1978). Looking at data from the World as a whole, a greater incidence amongst men than women has always been associated elsewhere with alcoholism (Harris, 1941).

Alcoholic pellagra does not seem to be associated with any particular basal diet. Before the routine fortification of cereals with a vitamin mix that included nicotinic acid, it was seen regularly amongst alcoholics in the Northern parts of the U.S.A. where wheat flour and potatoes were staples and corn was not (Harris, 1941). For the alcoholic pellagrins studied in

Brazil (Vannucchi et al., 1982) the diet was based on a roughly 3:1 mix of rice and Phaseolus vulgaris beans (H. Vannucchi, private communication).

There are obviously some remaining mysteries as to why pellagra occurs in some conditions and not in others. Why did it spring up in epidemic proportions in parts of the U.S.A. between 1906 and 1910 ? It has been argued elsewhere that one factor could be the development of the commercial degermination of corn meal at the beginning of this period (Lewin, 1980; Carpenter, 1981a). Also, although vitamin depletion is a necessary condition for the disease appearing, we do not understand why the appearance of it is stimulated by strong sunlight. This and other "stress" factors such as mycotoxins could still, for example, be factors determining whether the disease will appear at a dietary intake of 7.9 mg NE/1,000 kcal, or only at a level below 4.4 mg. So many dogmatic theories have already been overturned in the history of this subject that we should not be overconfident that we "know it all now".

REFERENCES

[The references marked "*" have been reproduced whole or in part, and in English translation where necessary, in Carpenter (1981b)]

Alsberg, C.L. (1909) Agricultural aspects of the pellagra problem in the United States. New York Med. J. 90:50-54.

*Aykroyd, W.R., Alexa, I. & Nitzulescu, J. (1935) Study of the alimentation of peasants in the pellagra area of moldavia (Rumania). Roum. Pathol. Exp. Microbiol. Arch. 8:407-422.

Bapurao, S. & Krishnaswamy, K. (1978) Vitamin B_6 nutritional status of pellagrins and their leucine tolerance. Am. J. Clin. Nutr. 31:819-824.

Carpenter, K.J. (1981a) Effects of different methods of processing maize on its pellagragenic activity. Fed. Proc. 40:1531-1535.

Carpenter, K.J. (1981b) Pellagra; Benchmark papers in the history of biochemistry II. Stroudsburg, PA. Dowden, Hutchinson & Ross.

Carpenter, K.J., Kodicek, E. & Wilson, P.W. (1960) The availability of bound nicotinic acid to the rat. 3. The effect of boiling maize in water. Br. J. Nutr. 14:25-34.

Carter, E.G.A. & Carpenter, K.J. (1981) Bound niacin in sorghum and its availability. Nutr. Res. 1:571-579.

Carter, E.G.A. & Carpenter, K.J. (1982a) The available niacin values of foods for rats and their relation to analytical values. J. Nutr. 112: 2091-2103.

Carter, E.G.A. & Carpenter, K.J. (1982b) The bioavailability for humans of bound niacin from wheat bran. Am. J. Clin. Nutr. 36:855-861.

Carvalho Da Silva, A., Fried, R. & De Angelis, R.C. (1952) The domestic cat as a laboratory animal for experimental nutrition studies. III. Niacin requirements and tryptophan metabolism. J. Nutr. 46:399-409.

Chaudhuri, D.K. & Kodicek, E. (1960) The availability of bound nicotinic acid to the rat. 4. The effect of treating wheat, rice and barley brans and a purified preparation of bound nicotinic acid with sodium hydroxide. Br. J. Nutr. 14:35-42.

*Christianson, D.D., Wall, J.S., Dimler, R.J. & Booth, A.N. (1968) Nutritionally unavailable niacin in corn. Isolation and biological activity. J. Agr. Food Chem. 16:100-104.

Cook, N. & Carpenter, K.J. (1982) Effects of leucine, tryptophan and pyridoxine on the niacin status of rats. Fed. Proc. 41:469 (abstract 1162).

Das, M.L. & Guha, B. (1960) Isolation and characterization of bound niacin (Niacinogen) in cereal grains. J. Biol. Chem. 235:2971-2976.

*Elvehjem, C.A., Madden, R.J., Strong, F.M. & Woolley, D.W. (1937) Relation of nicotinic acid and nicotinic acid amide to canine black tongue. J. Am. Chem. Soc. 59:1767-1768.

Finney, D.J. (1964) Statistical Methods in Biological Assays, 2nd ed., Charles Griffin and Co., Ltd, London, U.K.

Funk, C. (1912) The etiology of the deficiency diseases. J. State Medicine 20:341-368.

Ghosh, H.P., Sakar, P.K. & Guha, B.C. (1963) Distribution of the bound form of nicotinic acid in natural materials. J. Nutr. 79:451-453.

*Goldberger, J. & Tanner, W.F. (1922) Amino-acid deficiency probably the primary etiological factor in pellagra. Public Health Rep. 37:462-486.

Goldberger, J. & Tanner, W.F. (1925) A study of the pellagra-preventive action of dried beans, casein, dried milk and brewer's yeast with a consideration of the essential preventive factors involved. Public Health Rep. 40:54-79.

*Goldberger, J. & Wheeler, G.A. (1928) Experimental black tongue of dogs and its relation to pellagra. Public Health Rep. 43:179-216.

Goldberger, J., Waring, C.H. & Willets, D.G. (1915) The prevention of pellagra : a test of diet amongst institutional inmates. Public Health Rep. 30:3117-3131.

*Goldberger, J., Wheeler, G.A., Lillie, R.D. & Rogers, L.M. (1918a) A further study of experimental blacktongue with special reference to the blacktongue prevention in yeast. Public Health Rep. 43:657-694.

*Goldberger, J., Wheeler, G.A. & Sydenstricker, E. (1918b) A study of the diet of nonpellagrous and of pellagrous households, in textile mill communities in South Carolina in 1916. Am. Med. Assoc. J. 71:944-949.

Goldsmith, G.A., Gibbens, J., Unglaub, W.G. & Miller, O.N. (1956) Studies of niacin requirement in man. III. Comparative effects of diets containing lime-treated and untreated corn in the production of experimental pellagra. Am. J. Clin. Nutr. 4:151-160.

*Goldsmith, G.A., Miller, O.N. & Unglaub, W.G. (1961) Efficiency of tryptophan as a niacin precursor in man. J. Nutr. 73:172-176.

Goldsmith, G.A., Rosenthal, H.L., Gibbens, J. & Unglaub, W.G. (1955) Studies of niacin requirement in man. II. Requirement on wheat and corn diets low in tryptophan. J. Nutr. 56:371-386.

Gopalan, C. & Jaya Rao, K.V. (1975) Pellagra and amino acid unbalance. Vit. & Horm. 33:505-528.

Gopalan, C. & Srikantia, S.G. (1960) Leucine and pellagra. Lancet. I. 954-957.

Harper, A.E., Punekar, B.D. & Elvehjem, C.A. (1958) Effect of alkali treatment on the availability of niacin and amino acids in maize. J. Nutr. 66:163-172.

Harris, H.F. (1919) The Macmillan Co., New York.

Harris, S. (1941) Clinical Pellagra. C.V. Mosby Co., St.-Louis.

Heidelberger, C., Abraham, E.P. & Lepkovsky, S. (1948) Concerning the mechanism of the mammalian conversion of tryptophan into nicotinic acid. J. Biol. Chem. 176:1461-1462.

Horwitt, M.K. (1958) Niacin-tryptophan requirements of man. J. Am. Diet. Assoc. 34:914-919.

Horwitt, M.K., Harvey, C.C., Rothwell, W.S., Cutler, J.L. & Haffron, D. (1956) Tryptophan-niacin relationships in man. J. Nutr. (Suppl. 1) 1-43.

*Knight, B.C.J.G. (1937) The nutrition of Staphylococcus aureus; nicotinic acid in vitamin B₁. Biochem. J. 31:731-737.

*Knox, W.E. & Auerbach, V.H. (1955) The hormonal control of tryptophan peroxidase in the rat. J. Biol. Chem. 214:307-313.

Kodicek, E. (1940) Estimation of nicotinic acid in animal tissues, blood and certain foodstuffs. Biochem. J. 34:712-735.

Kodicek, E. & Pepper, C.R. (1948) Microbiological estimation of nicotinic acid and comparison with a chemical method. J. Gen. Microbiol. 2:306-314.

Kodicek, E. & Wilson, P.W. (1959) The availability of bound nicotinic acid to the rat. 1. The effect of lime-water treatment of maize and subsequent baking into tortilla. Br. J. Nutr. 13:418-431.

Krehl, W.A., Sarma, P.S., Teply, L.J. & Elvehjem, C.A. (1946) Factors affecting the dietary niacin and tryptophan requirement of the growing rat. J. Nutr. 31:85-106.

Krehl, W.A. & Strong, F.M. (1944) Studies on the distribution, properties and isolation of a naturally occurring precursor of nicotinic acid. J. Biol. Chem. 156:1-12.

*Krehl, W.A., Teply, L.J., Sarma, P.S. & Elvehjem, C.A. (1945) Growth-retarding effect of corn in nicotinic acid-low rations and its counteraction by tryptophan. Science 101:489-490.

Laguna, J. & Carpenter, K.J. (1951) Raw versus processed corn in niacin-deficient diets. J. Nutr. 45:21-28.

Lewin, W.J. (1980) Possible role of the degermination of corn in the development of pellagra. M.S. thesis. Department of Nutritional Sciences, Univ. of California, Berkeley.

Manson, J.A. & Carpenter, K.J. (1978) The effect of a high level of dietary leucine on the niacin status of dogs. J. Nutr. 108:1889-1898.

Mason, J.B. & Kodicek, E. (1973) The chemical nature of the bound nicotinic acid of wheat bran : studies of partial hydrolysis products. Cereal Chem. 50:637-646.

Mason, J.B. & Kodicek, E. (1970) The metabolism of niacytin in the rat : studies of excretion of nicotinic acid metabolites. Biochem. J. 120:509-514.

Mason, J.B., Gibson, N. & Kodicek, E. (1973) The chemical nature of the bound nicotinic acid of wheat bran : studies of nicotinic acid-containing macromolecules. Br. J. Nutr. 30:297-311.

Massieu, H., Cravioto, O.Y., Cravioto, R.O., Guzman, G.J. & de L. Surez Soto, M. (1956) New data concerning the effects of maize and tortilla on the growth of rats fed on diets low in tryptophan and niacin. Cienca (Mexico) 16:24-30.

May, J.M. & McLellan, D.L. (1972) The ecology of malnutrition in Mexico and Central America. New York: Hafner; pp. 103, 148 and 227.

McCollum, E.V. & Davis, M. (1915) The influence of certain vegetable fats on growth. J. Biol. Chem. 21:179-182.

McCollum, E.V. & Pitz, W. (1917) The "Vitamine" hypothesis and deficiency diseases : a study of experimental scurvy. J. Biol. Chem. 31:229-253.

McCollum, E.V., Simmonds, N. & Pearson, H.T. (1919) A biological analysis of pellagra-producing diets. VI. Observations on the faults of certain diets comparable to those employed by man in pellagrous districts.

Melnick. D., Hochberg, M. & Oser, B.L. (1945) Physiological availability of the vitamins. I. The human bioassay technic. J. Nutr. 30:67-79.

Nakagawa, I. & Sasaki, A. (1977) Effect of an excess intake of leucine with and without additions of vitamin B_6 and/or niacin, on tryptophan and niacin metabolism in rats. J. Nutr. Sci. Vitaminol. 23:535-548.

National Research Council (1980) Recommended Dietary Allowances, Revised 1979. Washington, D.C. : National Academy of Sciences.

*Nishizuka, Y. & Hayaishi, O. (1963) Enzymic synthesis of niacin nucleotides from 3-hydroxyanthranilic acid in mammalian liver. J. Biol. Chem. 238: PC483-PC484.

Patterson, J.I., Brown, R.R., Linksweiler, H. & Harper, A.E. (1980) Excretion of tryptophan-niacin metabolites by young men : effects of tryptophan, leucine and vitamin B_6 intakes. Am. J. Clin. Nutr. 33:2157-2167.

Pearson, W.N., Stempfel, S.J., Valenzuela, J.S., Utley, M.H. & Darby, W.J. (1957) The influence of cooked vs. raw maize on the growth of rats receiving a 9% casein ration. J.Nutr. 62:445-463.

Roussel, F.H. (1845) La Pellagre. Paris.

*Salas, I. (1863) Etiology and prophylaxis of pellagra. M.D. Thesis. Faculté de Médecine de Paris. 44:26-35.

*Sambon, L.W. (1910) Progress report on the investigation of pellagra. J. Trop. Med. Hyg. 13:289-300.

Sandhu, J.S. & Fraser, D.R. (1981) The metabolic origin of trigonelline in the rat. Biochem. J. 200:495-500.

Sato, P. & Udenfriend, S. (1978) Studies on ascorbic acid related to the genetic basis of scurvy. Vitam. & Horm. 36:33-52.

Schoental, R. (1980) Mouldy grain and the aetiology of pellagra : the role of toxic metabolites of Fusarium. Biochem. Soc. Trans. 8:147-150.

*Searcy, G.H. (1907) An epidemic of acute pellagra. Med. Ass. State of Alabama Trans., 387-392.

Sebrell, W.H. & Butler, R.E. (1938) Riboflavin deficiency in man. Publ. Health Rep. 53:2282-2284.

*Siler, J.F., Garrison, P.E. & MacNeal, W.J. (1914) The relation of methods of disposal of sewage to the spread of pellagra. Arch. Intern. Med. 14: 453-457, 469, 474.

Singal, S.A., Syndenstricker, V.P. & Littlejohn, J.M. (1948) The role of tryptophan in the nutrition of dogs on nicotinic acid-deficient diets. J. Biol. Chem. 176:1051-1062.

*Spies, T.D., Bean, W.B. & Ashe, W.F. (1939) Recent advantages in the treatment of pellagra and associated deficiencies. Ann. Intern. Med. 12:1830-1844.

Squibb, R.L., Braham, J.E., Arroyave, G. & Scrimshaw, N.S. (1959) A comparison of the effects of raw corn and tortillas (lime-treated corn) with niacin, tryptophan or beans on the growth and muscle niacin of rats. J. Nutr. 67:351-361.

Stiebeling, H.K. & Munsell, H.E. (1932) Food supply and pellagra incidence in 73 South Carolina farm families. Techn. Bull. No 333. U.S. Department of Agriculture, Washington, D.C.

*Thiery, F. (1932) Description of a malady called Mal de la Rosa. J. Méd., Chir. Pharm., Paris 2:337.

Vannucchi, H., Mello de Oliveira, J.A. & Dutra de Oliveira (1982) Tryptophan metabolism in alcoholic pellagra patients : measurements of urinary metabolites and histochemical studies of related muscle enzymes. Am. J. Clin. Nutr. 35:1368-1374.

Warburg, O. & Christian, W. (1935) A co-enzyme problem. Biochem. Z. 275:464.

IRON REQUIREMENTS AND BIOAVAILABILITY OF DIETARY IRON

Leif HALLBERG

University of Göteborg, Sahlgren's Hospital, Department of
Medecine II, Göteborg, Sweden

SUMMARY

Over the past few decades so much knowledge has been gained
about iron needs, dietary iron availability and adequacy that it
is now one of the best defined nutrients in these respects.
Development of new, accurate methods for the measurement of both
the losses of iron from the body and the absorption of iron from
the diet has significantly contributed to this situation.

Present knowledge of iron needs is summarized. Specific to
iron are the much higher needs in women than in men and the
great variation in needs between different women due to a marked
physiological variation in menstrual iron losses and to the
effects of pregnancies.

Iron availability is discussed separately for heme and non-
heme iron (the major type of food iron). Heme iron in small
amounts is, on average, better absorbed than non-heme iron. The
absorption of heme iron is influenced very little by the iron
status of the subject and by the other food components in the
diet with the exception of meat which stimulates absorption. On

the other hand, the absorption of non-heme iron is markedly
influenced both by the iron status of the subject and a great
number of dietary factors.

The absorption of iron from the diet is thus determined more
by meal composition than by the amount of iron present in the
diet. The great variation in absorption between different meals
is illustrated and the importance of various factors influencing
non-heme iron absorption is also demonstrated. Whilst the mode
of food preparation itself influences iron absorption, meat or
fish and ascorbic acid are some principal food constituents that
enhance absorption of iron. On the other hand, several factors
like tannins, phytates, phosphates, soya protein products and
various "dietary" fibres have been reported to inhibit non-heme
iron absorption.

A consideration of the nutritional adequacy of iron high-
lights the importance of methods of evaluation, particularly the
usefulness of the "bioavailable nutrient density" (BND) approach
for different meals; BND for iron represents the amount of iron
absorbed per 1'000 kcal (4'180 kJ).

The main problem in iron nutrition in Western countries today
is that arising from the combination of a low-energy intake,
especially in women having the highest iron needs, with a
conservatism in the choice of meals/meal composition. Provision
of bioavailable dietary iron to meet needs has hence not been
adjusted to "modern life".

* * *

Over the past few decades much knowledge has been gained about iron
needs, dietary iron availability and adequacy. Iron is one of the nutrients
which are best defined in these respects. The reason is that new accurate
methods have been developed for the measurement of both the losses of iron
from the body and the absorption of iron from the diet. My assessment of

iron nutrition will be limited to adults as a forthcoming Nestlé symposium on iron will be devoted to infants and children.

The paper consists of three parts : (1) a summary of present knowledge of iron requirements in adults, (2) a discussion of recent information about the bioavailability of dietary iron, and (3) an analysis of methods to evaluate the adequacy of a diet with respect to iron.

Iron requirements

The total amount of iron in the body is about 2000-3000 mg in adults. Since iron is continuously recycled in the body, the daily iron requirements are relatively small and consist of the amounts of iron lost from the body and the amounts necessary for growth and development.

Iron is regularly lost from the body by loss of desquamated cells from external and internal body surfaces such as skin, gastrointestinal tract and urinary tract. Iron is also lost in all subjects by minor losses of blood. These basal physiological losses amount to 14 µg of iron/kg body weight/ day in healthy men and women (e.g. 0.8 mg/d in a 55 kg woman and 1.0 mg in a 70 kg man). For a review see Bothwell et al., 1979.

In women of childbearing age, the menstrual iron losses have to be added to the basal losses. Of great importance in understanding the critical iron balance situation in a great proportion of women is the fact that the menstrual iron losses are very constant in the particular woman but vary markedly between different women. This explains why some women have much greater iron requirements than others. In a large group of women, randomly selected from the general population, menstrual blood loss resulted in an average daily iron loss of 0.6 mg (Hallberg et al., 1966). In 25% of these women the losses exceeded 0.9 mg/day and in 10% they exceeded 1.4 mg/day. Thus, the total iron requirements in 10% of menstruating women exceed 2.2 mg/day.

Studies in twins indicate that the magnitude of the menstrual iron losses is to a large extent genetically controlled (Rybo & Hallberg, 1966). Another

interesting fact is the observation that the frequency distribution curve of
the menstrual iron losses and the average losses per unit body weight are
about the same in Burmese women as in women in Sweden, England and Canada.
These observations imply that the iron requirements in our ancestors were
probably very similar to the requirements in present-day women. Exceptions
may be due to modern anticonceptional methods - oral contraceptives reduce
menstrual losses by half and present intrauterine contraceptive devices
increase the losses by about 50%.

The iron requirements of _pregnancy_ are also well defined. The total iron
requirements are about 1000 mg. Iron is needed to cover the basal losses of
the mother (240 mg), the increase in maternal red cell mass (about 500 mg),
and the requirements of placenta and fetus (about 300 mg).

A great problem is that these enhanced requirements are not evenly
distributed over the whole period of pregnancy, but develop chiefly in the
latter half or third of pregnancy, because of the exponential growth of the
fetus. In the last month of pregnancy for instance, the total iron requirements may amount to 7-8 mg/day (Fig. 1).

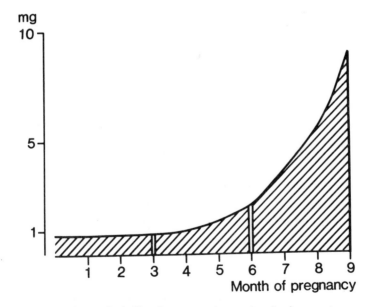

Fig. 1 : Estimated total daily iron requirements during pregnancy.

Such requirements can never be met from diet alone and it is reasonable to assume that a main function of iron stores in the body is to counteract the development of iron deficiency in the later part of pregnancy. There is no increase in the absorption of iron from the diet during early pregnancy when the iron requirements are low but absorption increases during the latter part of pregnancy. It can be calculated that iron stores of about 500 mg are needed to balance the iron requirements during pregnancy. Iron stores of this magnitude are rarely found today in young women and this explains why it is recommended to give iron in tablet form during the latter part of pregnancy. This seemingly irrational and unphysiological approach results in industrialized countries from the present low dietary intake of energy and iron in women and in most developing countries from the low bioavailability of iron from present diets.

During the lactation period there are usually no menstrual iron losses and the daily iron requirements are then composed of the basal losses of 0.8 mg and the iron lost into breast milk (about 0.25 mg/d), or a total of about 1 mg/d. It is probable that in early man iron stores were replenished to adequate levels during the lactation period.

Bioavailability of dietary iron

There are two kinds of iron in the diet with respect to the mechanism of absorption - heme iron (derived mainly from hemoglobin and to a minor extent from myoglobin) and non-heme iron (derived mainly from cereals, vegetables and fruits). Heme iron is taken up into the mucosal cells as an iron-porphyrin complex which is split in these cells by a specific enzyme. Non-heme iron is probably taken up by a transferrin-like receptor on the surface of the mucosal cells. The absorption of these two kinds of iron is influenced differently by dietary factors and by the iron status of the subjects (Hallberg et al., 1979).

Heme iron forms only a minor part of the iron intake - up to 10-15%. Its absorption is influenced by meat in the diet but not by other dietary factors. There is little if any influence of the subject's iron status on heme iron absorption. The absorption of non-heme iron, on the other hand, is markedly influenced by a great number of known and unknown factors in the diet and by the iron status of the subjects.

Most knowledge about the absorption of iron from the diet has been obtained during the last 10 years. The reason is that new methodology was introduced at that time which made it possible for the first time to measure the absorption of iron from meals (Hallberg and Björn-Rasmussen, 1972; Cook et al., 1972). The basis for the new method was the unexpected observation that an inorganic radioiron tracer added to different foods rapidly and uniformly labelled the various non-heme iron compounds in a meal, probably by an isotopic exchange. Heme iron was not labelled by an inorganic radio-iron tracer and the concept of two separate pools of iron in the gastro-intestinal tract was introduced - the heme iron and the non-heme iron pools. The use of two different radioiron isotopes enables these two pools to be independently and simultaneously labelled with biosynthetically radioiron-labelled hemoglobin and with an extrinsic, inorganic radioiron tracer.

Non-heme iron is the main source of dietary iron and the bioavailability of this iron under different conditions is, therefore, of great importance for iron nutrition. A main problem in studies on the absorption of non-heme iron is the marked effect of the iron status of the subjects - more iron is absorbed by iron-deficient and less by the iron-replete subjects - leading to a marked subject-to-subject variability. This makes it difficult to determine whether differences in absorption between meals studied in different groups of subjects relate to properties of the meals or to the iron status of the subjects. This problem is overcome, however, by adjusting in each subject the absorption from a meal to the absorption from a standard dose of inorganic radioiron given at physiological levels under standard-ized conditions. In this way an independent measure of the absorptive capacity for iron is obtained in each subject (Layrisse et al., 1969).

An informal agreement has been reached among workers in different countries to use 3 mg of iron as ferrous "ascorbate" for this reference dose. Subjects who are borderline iron-deficient absorb about 40% from this dose and it has, therefore, been suggested that the absorption from meals be normalized to correspond to an absorption of 40% from the reference doses.

In this way, the bioavailability of iron from a meal, studied in a group of subjects, can be expressed as a single figure corresponding to the iron absorption in subjects who have borderline iron-deficiency, i.e. subjects with absent iron stores but who have no anaemia (Magnusson et al., 1981).

Factors affecting the bioavailability of dietary non-heme iron

As mentioned earlier the non-heme iron in most foods is completely and rapidly labelled by an added inorganic radioiron tracer and this implies that an isotopic exchange takes place. Moreover, it suggests that all non-heme iron compounds present in a meal can be regarded as part of a common pool. The properties of this pool and thus the absorption from the pool, i.e. the meal, can be markedly affected by a number of factors in the diet - some inhibiting and some enhancing the absorption. For a review see Hallberg, 1981a. The most potent factors increasing the absorption are ascorbic acid and meat/fish. Factors inhibiting the absorption are for example tannins (in tea, vegetables and certain cereals) and bran, probably because of its content of phytates (see below). Quantitative information of the effect of various factors is easily obtained by serving a meal, with or without the factor to be studied on alternate days to the same group of subjects, labelling the meals with two different radioiron isotopes.

The markedly enhancing effect of meat and fish on iron absorption was early observed by Layrisse et al., 1968. The mechanism of action is still not settled. It is of interest, however, that the enhancing effect is of about the same magnitude for heme and non-heme iron (Hallberg et al., 1979). The low bioavailability of iron in diets in developing countries was long considered to be due solely to their low content of meat and fish; however, this is not true. The absorption of non-heme iron from meals with a high content of ascorbic acid can actually be as good as or even better than from the same basal meal to which meat is added to enhance the absorption. Meat and ascorbic acid probably enhance non-heme iron absorption by independent mechanisms and the absorption-promoting effects will, therefore, be additive.

Ascorbic acid seems to be the most potent enhancer of non-heme absorption. The extent of enhancement depends on the amount of ascorbic acid and

on the initial bioavailability of iron in the meal. Addition of 50 mg
ascorbic acid to a simple maize meal will increase the absorption about
3-5 times, whereas the absorption increase from a meal with a high basal
bioavailability may be 2-3 fold. The increase in bioavailability in absolute
amounts, however, will be considerable in both cases. A glass of orange
juice containing 70 mg ascorbic acid increased the iron absorption from a
hamburger meal 2.2 times and a mixed vegetable salad composed of lettuce,
green pepper and tomatoes containing 45 mg ascorbic acid increased the
absorption 1.8 times from the same hamburger meal. For a review on the
effect of ascorbic acid on the bioavailability of iron from food, see
Hallberg, 1981b.

Tannins are the probable cause of the marked reduction in iron absorption
observed when meals are served with tea. This was first reported by Disler
et al., 1975. Tea served with a continental breakfast reduces the absorption
to less than half (Rossander et al., 1979). The effect is about the same
when tea is served with a hamburger meal (Hallberg & Rossander, 1982c).
Vegetables and cereals with a high content of tannins also inhibit the non-
heme iron absorption. This is an important area for further research.

Soy protein products have been reported to inhibit non-heme iron absorption (Cook et al., 1981). Other workers have obtained different results
(Fig. 2) (Hallberg & Rossander, 1982d). The divergencies in results, however, are more apparent than real and are mainly due to the interpretation
of the results. When soy is added to a meat-containing meal to make a bigger
hamburger (meat extension), the percentage of non-heme iron absorbed is
usually decreased but the actual amount of iron absorbed is increased as the
added soy product has a high iron content. However, when soy is used to
replace a portion of the meat in a hamburger (meat substitution), there is a
decrease in the total amount of iron absorbed from the meal. The reduction
is partly due to a decrease in the enhancing effect of meat on non-heme iron
absorption and partly to a decrease in heme iron content of the meal
(Fig. 3).

The overall effect of soy protein on iron nutrition depends on several
factors such as the iron status of the population, the amount and type of
soy protein used and to what extent it is used as a meat substitute.

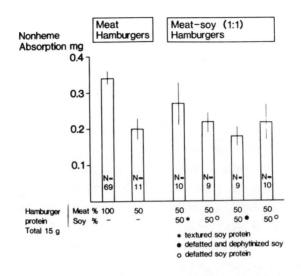

Fig. 2 : Non-heme iron absorption from a hamburger meal. The meat content had either been reduced by half or exchanged for different forms of soy protein.

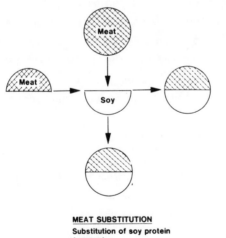

Fig. 3 : Two interpretations of the effect of soy protein on non-heme iron absorption from hamburgers.

In populations with a low meat intake, where soy is added to the diet to increase the protein intake, the iron absorption is significantly increased by soy, mainly because of its high iron content.

Wheat bran has been shown markedly to inhibit non-heme iron absorption. The earliest observations were made in balance studies a long time ago by Widdowson & McCance (1942) and these results have been confirmed and extended in studies using the extrinsic tag method (Björn-Rasmussen, 1974). The relative role of phytates and fibres in explaining the inhibition of the non-heme iron absorption by bran is not yet settled as divergent results have been found. One group claims that iron in bran is present as monoferric phytate which is as well absorbed as ferric chloride. Moreover, an inhibitory effect of bran was also seen after degradation of its phytate by endogenous phytase (Simpson et al., 1981). In recent studies in our laboratory (Hallberg & Rossander, unpublished) divergent results were obtained indicating that the phytates are the main factor responsible for the inhibiting effect of bran on iron absorption. As shown in Fig. 4, the addition of bran to a wheat bread markedly reduced the absorption. Washing the bran with water did not eliminate its inhibiting effect whereas washing with dilute hydrochlorid acid increased the absorption to the same level as from bread baked of white wheat flour. Restitution of the phytates by the addition of sodium phytate again reduced the absorption. It can be argued (1) that hydrochloric acid extracts not only phytates but also some other fibre fraction inhibiting iron absorption and (2) that sodium phytate is an unphysiological compound, as native phytates are in the form of potassium and magnesium salts and the iron in the form of monoferric-phytate.

In further studies, we have therefore compared the absorption from white wheat bread to which we added either bran or a "physiological" mixture of monoferric phytate and magnesium - potassium phytates in amounts corresponding to the contents of both iron and phytates in the bran. The difference between the two kinds of bread in this study is thus only in the fibre fractions of bran. The two kind of breads were labelled with different radioiron isotopes and served to the same subjects on alternate days.

It is evident from Fig. 5 that the absorption was the same when adding bran and the phytate mixture, indicating that the bran fibre as such had no significant, additional, inhibitory effect on non-heme absorption.

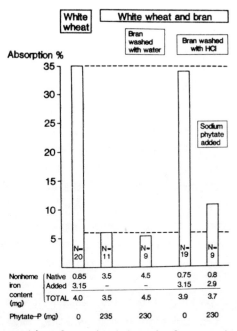

Nonheme iron content (mg)		White wheat	Bran washed with water	Bran washed with HCl (Sodium phytate added)	
	Native	0.85	3.5	4.5	0.75	0.8
	Added	3.15	–	–	3.15	2.9
	TOTAL	4.0	3.5	4.5	3.9	3.7
Phytate–P (mg)		0	235	230	0	230

Fig. 4 : Iron absorption from wheat bread after various treatments of the wheat bran.

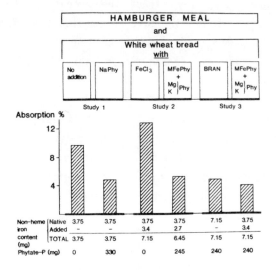

Non-heme iron content (mg)		No addition	Na Phy	FeCl₃	MFePhy + Mg K Phy	BRAN	MFePhy + Mg K Phy
	Native	3.75	3.75	3.75	3.75	7.15	3.75
	Added	–	–	–	3.4	2.7	–	3.4
	TOTAL	3.75	3.75	7.15	6.45	7.15	7.15
Phytate–P (mg)		0	330	0	245	240	240

Fig. 5 : Iron absorption from wheat bread to which was added bran or a mixture of monoferric phytate and magnesium potassium phytate corresponding to the assumed naturally occurring phytates in bran. In these studies, the bread was given together with a hamburger meal.

In another experiment, the iron absorption from white bread was compared with the same bread to which the monoferric phytate/magnesium-potassium phytate mixture was added. The reduction in iron absorption was about the same as when bran was added.

A probable explanation of why our results are not consistent with the ones reported from the group in Kansas City may be that the latter group only added monoferric phytate to the white wheat flour and not the further 90% of phytates (as magnesium and potassium phytates) needed to make it equivalent to the actual phytate content of the bran.

The effect of various fibres on iron absorption has been very much discussed in recent years. Very few studies seem to have been made in man, however, and most reports deals with in vitro studies or animal studies.

Fibres may, theoretically, affect iron absorption in several ways : they may act as ion-exchangers and bind iron ions, they may decrease the rate of diffusion from the intestinal content to the mucosal surface, they may affect the rate of emptying of the stomach and/or the rate of transport along the intestines, they may affect the secretion of bile, pancreatic juice etc. In summary, it is difficult to predict the effect of fibres on iron absorption. As judged from reports on the effect of fibres on other nutrients such as lipids and carbohydrates, the effect on iron absorption might be quite marked.

Most of our studies on the effect of fibres have been made by baking white wheat rolls containing 75 g wheat flour and 5 g of various fibre materials. Pectins might inhibit iron absorption by increasing the viscosity of the intestinal content or by acting as ion-exchangers. We found, however, that pectins, both with a high and a low degree of methoxylation, did not affect non-heme iron absorption.

Guar-gum, which has been shown markedly to reduce the rate of glucose absorption and the rate of insulin release was found to increase significantly ($p < 0.01$) non-heme iron absorption.

Ispagula is a plant mucin much used in bulk laxatives. A slight reduction in the non-heme iron absorption was seen ($p < 0.05$). A comparison of the effect of various fibres materials on the non-heme absorption is shown in Fig. 6.

The effects of various cereals and vegetables on non-heme iron absorption are probably not only related to phytates, fibres and tannates. In a comparison of iron absorption from meals containing different cereals as the main source of energy, it was evident that iron is better absorbed from meals with wheat than with rice and maize (in this order) (Fig. 7).

A comparison of non-heme iron absorption was, therefore, made from rolls only, baked with flour from wheat and rice. The relative superiority of wheat was the same. Rolls were then baked of starch only. The absorption from rolls baked of rice and maize starches was significantly lower than from wheat starch rolls. The relatively higher absorption from maize starch especially compared with the ratio in whole meals, can be explained by the absence of phytates in the maize starch rolls (Fig. 7).

These results show that different starches affect iron absorption differently. A probable reason may be that iron is partially bound to starches and that this binding is of variable strength or that the binding to starches is the same, but that the rate of digestion is different for different starches, in this way releasing iron at different rates in the intestines. This latter interpretation fits with recent observations of a significantly faster absorption of glucose from wheat flour than from rice flour. It is tempting to speculate that part of the effects of different plant materials in nutrition, today claimed to be due to various fibres, may actually be due to different properties of starches in different plants.

The examples given of the effects of various factors on non-heme iron absorption, illustrate the complexity of the chemistry of food iron absorption and the difficulty to predict the bioavailabiliy of iron from different meals.

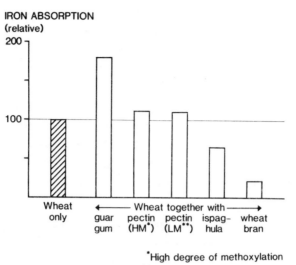

*High degree of methoxylation
**Low degree of methoxylation

Fig. 6 : Effect of different forms of fibres on non-heme iron absorption from rolls baked of white wheat flour (75 g) to which 5 g of fibre was added. The pectins had either a high (HM) or low (LM) degree of methoxylation.

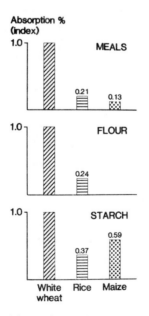

Fig. 7 : Relative absorption of non-heme iron from meals composed mainly of white wheat, rice or maize (top part of figure); from rolls prepared from wheat flour and rice flour; and from rolls prepared from starch from wheat, rice and maize (bottom part of figure).

Certain factors such as ascorbic acid, meat, fish, tannates and phytates have the most marked effects. The contents of these factors can usually explain a great of part but not the whole variation in bioavailability of non-heme iron between meals.

Fig. 8-11 illustrate the overall effects of various factors on bioavailability on non-heme iron in meals. Fig. 8 shows the variation in absorption between different Western-type breakfast meals (Rossander et al., 1979). Note the marked effects of tea-inhibiting and of orange juice-stimulating the non-heme iron absorption. In total, there was a 6-fold variation in iron absorption. Fig. 9 shows the variation in absorption from 13 Western-type lunch-dinner meals (Hallberg & Rossander, 1982a). Note the several-fold variation in bioavailability. The percentage of non-heme iron absorbed varied from 3% (for pancakes) to 45% (for a meal composed of sauerkraut and sausage). The reasons for this latter very high absorption is not yet clear. Fig. 10 shows a 5-6 fold variation in absorption of non-heme iron from 10 different whole meals with a similar energy content (about 1'000 kcal.) (Hallberg & Rossander, 1982b). Fig. 11 shows the absorption from two vegetarian meals with the same energy, protein and iron content. One of the meals had a high content of ascorbic acid due to cauliflower (74 mg). The iron absorption from this meal was of the same magnitude as from meals with a high meat content. The addition and substraction of different food items showed that besides the high content of ascorbic acid, other factors must also be responsible for the differences in bioavailability of iron in these two vegetarian meals, e.g. the content of phytates.

Assessment of the nutritive adequacy of a diet with respect to iron

Several questions should be answered. For example :

1. How much iron is absorbed from the diet ?

2. How much iron is needed; i.e. what is the frequency distribution of the iron requirements ?

3. What proportion of the population risks an insufficient supply of iron and who are at risk ?

Basal breakfast with different drinks
coffea
milkchocolate
tea
Basal breakfast with coffea
no addition
and cornflakes + milk
one egg
one egg + bacon
orange juice
Basal breakfast with tea
no addition
and one egg + bacon
orange juice
0
0.10
0.20
0.30
0.40
Absorption (mg)

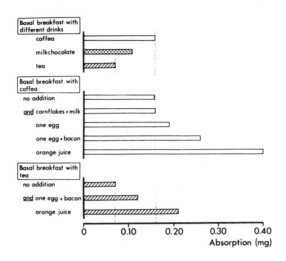

Fig. 8 : Non-heme iron absorption from different Western-type breakfast meals.

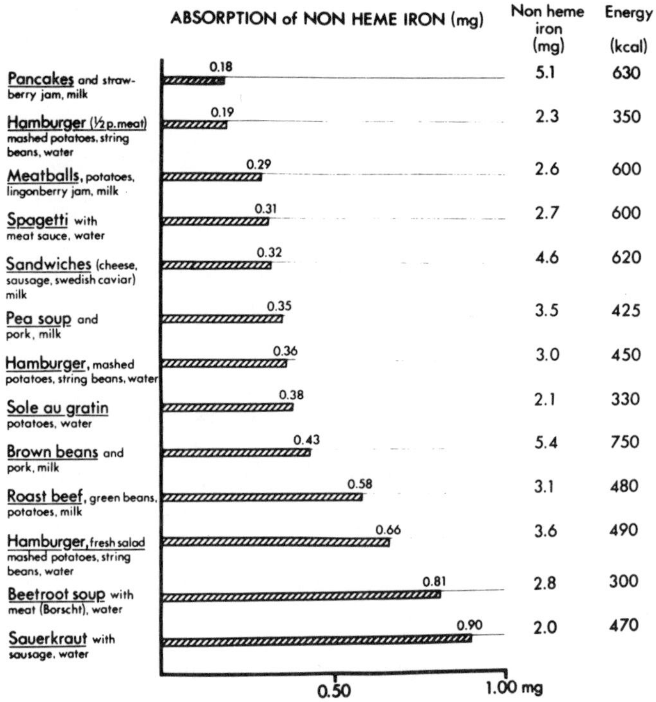

ABSORPTION of NON HEME IRON (mg)
Non heme iron (mg)
Energy (kcal)
Pancakes and strawberry jam, milk
0.18
5.1
630
Hamburger (½ p. meat) mashed potatoes, string beans, water
0.19
2.3
350
Meatballs, potatoes, lingonberry jam, milk
0.29
2.6
600
Spagetti with meat sauce, water
0.31
2.7
600
Sandwiches (cheese, sausage, swedish caviar) milk
0.32
4.6
620
Pea soup and pork, milk
0.35
3.5
425
Hamburger, mashed potatoes, string beans, water
0.36
3.0
450
Sole au gratin potatoes, water
0.38
2.1
330
Brown beans and pork, milk
0.43
5.4
750
Roast beef, green beans, potatoes, milk
0.58
3.1
480
Hamburger, fresh salad mashed potatoes, string beans, water
0.66
3.6
490
Beetroot soup with meat (Borscht), water
0.81
2.8
300
Sauerkraut with sausage, water
0.90
2.0
470
0.50
1.00 mg

Fig. 9 : Absorption of non-heme iron from 13 different meals.

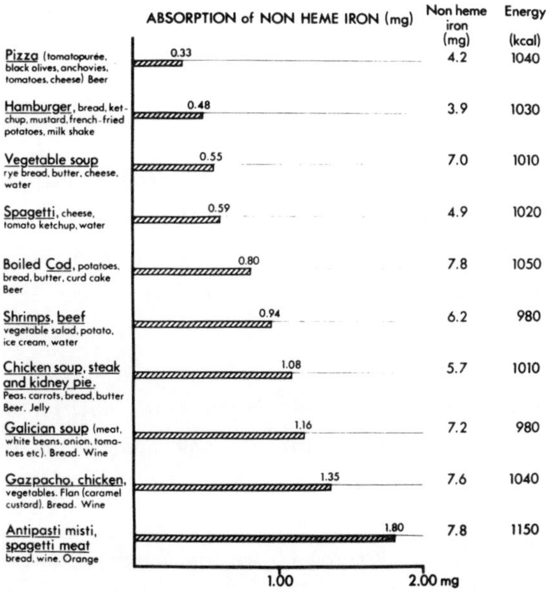

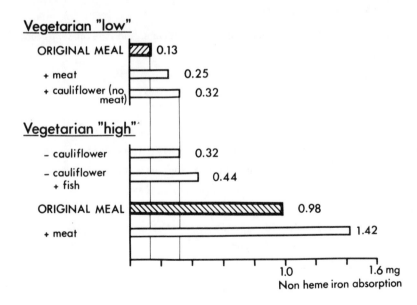

Fig. 10 : Absorption of non-heme iron from 10 different whole meals with a similar energy content.

Fig. 11 : Absorption of non-heme iron from two vegetarian meals with the same iron, protein and energy content, and the effect of subtracting or adding various food components.

The marked variation in iron requirements and the marked variation in bioavailability of dietary iron may make it impossible to outline a meaningful assessment. The following method, however, may be useful.

1. Women of childbearing age have the highest iron requirements. If a diet is considered as adequate when it covers the requirements of iron in 90% of these women (disregarding pregnancies) it can be calculated that 2.2 mg iron/d need to be absorbed to maintain them in an iron replete state.

2. A limiting factor for an adequate supply of nutrients is the energy expenditure and thus the energy intake. The absorption of iron per unit of energy consumed is thus critical.

3. The amount of iron absorbed from a diet depends not only on its contents of heme and non-heme iron and on the amounts of various factors in the diet affecting the bioavailability but also on the composition of individual meals, for instance how much ascorbic acid is present in meals with the highest iron content.

   It is thus necessary to know the meal pattern and the absorption of iron from individual meals.

4. A measure of the nutritive values of a meal will be obtained if the amount of a nutrient (iron) <u>absorbed</u> is related to the energy content (Hallberg, 1981c). This expression, the bioavailable nutrient density (BND), must be calculated for all main meals. The average iron absorption can then be estimated. A rough calculation of a typical Swedish weekly or monthly menu for women gave an average bioavailable nutrient density (BND) of 0.9. The heme iron absorption must of course also be included in the calculations. It should be mentioned that the variation in BND in the meals shown was 12-fold. Different meal patterns in different families, different school lunch programs, different ethnic and socio-economic groups etc. can be evaluated using the present method.

The assessment of the nutritive adequacy of a diet with respect to iron must thus be based on a comparison of (1) iron requirements, (2) energy intake and (3) bioavailable nutrient density. The higher the energy intake the lower the required BND (Fig. 12).

For men, the iron requirements are set to 1.0 mg and for 90% of women to cover requirements, to 2.2 mg; in 75%, 1.8 mg; and 50%, 1.4 mg Fe/d (the curved lines in Fig. 12). A horizontal line is drawn at a bioavailable nutrient density of 0.9 mg Fe/1'000 kcal. To cover the iron requirements in 90% of women from the diet, 2'500 kcal. of such a diet have to be consumed. The graph can also be used the other way. If the daily energy intake is as low as 2'000 kcal., a diet with a BND of 0.9 can only balance the iron requirements in about 80% of non-pregnant women of childbearing age. The data obtained correspond fairly well with the actual situation in our female population.

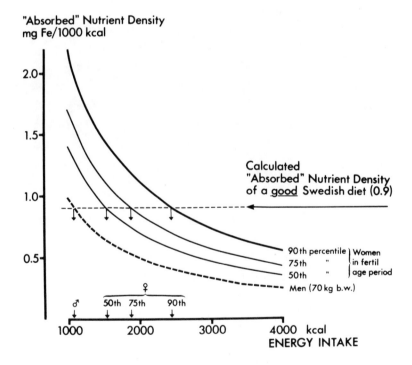

Fig. 12 : Bioavailable nutrient density needed at different energy intakes to cover iron requirements in men and in certain proportions of menstruating women. The estimated bioavailable nutrient density of a good Swedish diet today (0.9) is drawn as a dotted horizontal line. The energy intake needed to cover the iron requirements can then be read on the abscissa.

The figure 0.9 for BND is not precise and its variation is not studied. The figure is given only to illustrate its use in the present method of assessment. The present model can, of course, be applied in the evaluation of any diet and any nutrient, provided the iron requirements and the average BND are known.

CONCLUSIONS

The present analysis shows that methods are now available to make assessments of the adequacy of different diets with respect to iron. It is also possible to make calculations of the effects expected of different iron fortification programs using the same methodological principles and adding information about the bioavailability of the iron compounds used and about the extent to which the fortificant reaches different segments of the population (Hallberg, 1982).

The present analysis shows that the main problem in iron nutrition in Western countries today is the combination of a low energy intake, especially in women having the highest iron needs, with a conservatism in the choice of meals; i.e. in meal composition. The dietary iron bioavailability has thus been adjusted to "modern life".

ACKNOWLEDGEMENTS

The present paper was supported in part by a grant from the Swedish Medical Research Council MFR Project no B 80-19X-04721-05C.

REFERENCES

Björn-Rasmussen, E. (1974) Iron absorption from wheat bread. Influence of various amounts of bran. Nutr. Metab. 16:101-110.

Bothwell, T.H., Charlton, R.W., Cook, J.D. & Finch, C.A. (1979) Iron Metabolism in Man. Blackwell Scientific Publications, Oxford.

Cook, J.D., Layrisse, M., Martinez-Torres C., Walker, R., Monsen, E. & Finch, C.A. (1972) Food iron absorption measured by an extrinsic tag. J. Clin. Invest. 51:805-815.

Cook, J.D., Morck, I.A. & Lynch, S.R. (1981) The inhibitory effect of soy products on non-heme iron absorption in man. Am. J. Clin. Nutr. 34:2622-2629.

Disler, P.B., Lynch, S.R., Charlton, R.W., Torrance, J.D. & Bothwell, T.H. (1975) The effect of tea on iron absorption. Gut 16:193-200.

Hallberg, L. (1981a) Bioavailability of dietary iron in man. Annu. Rev. Nutr. 1:123-147.

Hallberg, L. (1981b) Effect of vitamin C on the bioavailability of iron from food. In : Vitamin C (ascorbic acid) (Ed. Counsell, J.N., Hornig, D.H.) Applied Science Publishers Ltd., London and New Jersey 1981, Chapter 3, pp. 49-61.

Hallberg, L. (1981c) Bioavailable nutrient density : a new concept applied in the interpretation of food iron absorption data. Am. J. Clin. Nutr. 34:2242-2247.

Hallberg, L. (1982) Iron nutrition and food iron fortification. Seminars in Hematology, vol. 19, No 1:31-41.

Hallberg, L. & Björn-Rasmussen, E. (1972) Determination of iron absorption from whole diet. A new two-pool model using two radioiron isotopes given as haem and non-haem iron. Scand. J. Haematol. 9:193-197.

Hallberg, L., Björn-Rasmussen, E., Howard, L. & Rossander, L. (1979) Dietary heme iron absorption. A discussion of possible mechanisms for the absorption-promoting effect of meat and for the regulation of iron absorption. Scand. J. Gastroenterol. 14:769-779.

Hallberg, L., Högdahl, A.M., Nilsson, L. & Rybo, G. (1966) Menstrual blood loss of population study. Acta Obstet. Gynecol. Scand. 45:25-56.

Hallberg, L. & Rossander, L. (1982a) Absorption of iron from Western-type lunch and dinner meals. Am. J. Clin. Nutr. 35:502-509.

Hallberg, L. & Rossander, L. (1982b) Bioavailability of iron from Western-type whole meals. Scand. J. Gastroenterol. 17:151-160.

Hallberg, L. & Rossander, L. (1982c) Effect of different drinks on the absorption of non-heme iron from composite meals. Hum. Nutr. Appl. Nutr. 36 A:116-123.

Hallberg, L. & Rossander, L. (1982d) Effect of soy protein on nonheme iron absorption in man. Am. J. Clin. Nutr. 36:514-520.

Layrisse, M., Cook, J.D., Martinez, C., Roche, M., Kuhn, I.N., Walter, R.B. & Finch, C.A. (1969) Food iron absorption : a comparison of vegetable and animal foods. Blood 33:430-433.

Layrisse, M., Martinez-Torres, C. & Roche, M. (1968) The effect of interaction of various foods on iron absorption. Am. J. Clin. Nutr. 21:1175-1183.

Magnusson, B., Björn-Rasmussen, E., Hallberg, L. & Rossander, L. (1981) Scand. J. Haematol. 27:201-208.

Rossander, L., Hallberg, L. & Björn-Rasmussen, E. (1979) Absorption of iron from breakfast meals. Am. J. Clin. Nutr. 32:2484:2489.

Rybo, G. & Hallberg, L. (1966) Influence of heredity and environment on normal menstrual blood loss. Acta Obstet. Gynec. Scand. 45:57-78.

Simpson, K.M., Morris, E.R. & Cook, J.D. (1981) The inhibitory effect of bran on iron absorption in man. Am. J. Clin. Nutr. 34:1469-1478.

Widdowson, E.M. & McCance, R.A. (1942) Iron exchanges of adults on white and brown bread diets. Lancet 1:588-591.

VITAMIN DEFICIENCIES IN RICE-EATING POPULATIONS
EFFECTS OF B-VITAMIN SUPPLEMENTS

Mahtab S. BAMJI

National Institute of Nutrition, Indian Council of Medical Research
Hyderabad, India

SUMMARY

Rice is the staple food in many countries of Asia. Recent nutrition surveys in eight states, conducted by the National Nutrition Monitoring Bureau of India, show that though the average energy intake is adequate, more than 50% of the households surveyed consumed less than the Recommended Dietary Allowance (RDA) of energy. These households generally had per capita incomes of less than Rupees 2/- (US$ 0.25) per day. The average intake of vitamin A was only 42% of the RDA and that of riboflavin, 70% of the RDA. The average intake of other nutrients such as thiamin, niacin, ascorbic acid, iron and calcium was adequate, although thiamin deficiency was present in populations where rice was the main cereal, but not in populations that consumed mixed cereal or cereal-millet diets. The magnitude of the riboflavin deficiency (after correction for energy) was also more marked in the former. Vitamin A intake was not related to the type of cereal, but had some relationship to the quantity of vegetables consumed.

Nutrition surveys from Japan also reveal deficiencies in intake of energy, vitamin A, thiamin and riboflavin. The Japa-

nese diet tends to be deficient by 20% in vitamin A and riboflavin, but not thiamin. Thus, vitamin A, riboflavin and energy
(in that order) are the major nutritional constraints in rice-
eating populations.

Clear-cut correlations between the magnitude of dietary
deficiency and the prevalence of signs and symptoms of vitamin
deficiency were not apparent in the comparisons between populations, suggesting that as well as dietary deficiency other
environmental factors play a role in the development of clinical
deficiency. Attempts to correlate clinical deficiency with the
magnitude of biochemical deficiency have also failed.

Recent studies aimed at examining the effects of food supplements (rural Gambian women) or vitamin supplements (rural Indian
boys) on vitamin status suggest that in some communities,
vitamin intakes close to the RDA fail to saturate the tissues,
as judged by biochemical tests. In the Indian boys, there was a
marked rise in urinary excretion of riboflavin during winter
when the incidence of respiratory infections was high. Metabolic
losses of vitamins due to infections may preclude tissue
saturation despite adequate dietary intake. Administration for
1 year of B-vitamins at levels close to the RDA failed to reduce
the prevalence of clinical deficiency signs, but did produce
some improvement in hand steadiness - a psychomotor test. The
data suggest that dietary improvement with regard to vitamins
may fail to produce the desired impact on the health of some
rural communities.

* * *

Over the last 100 years, nutrition scientists have tried to obtain
answers to questions such as what, why and how much one should eat, what
happens if one eats more or less than the required quantity, what are the
non-nutritional factors which influence nutrition status, and what should be
the strategies for achieving man's nutritional goals. While the question as

to what and why one should eat has been answered, that of nutrient requirements is still in the melting pot, because it is linked to the basic problem of definition of good health and detection of marginal malnutrition.

Fig. 1 describes the step-wise development of a deficiency disease. Tissue depletion can occur following inadequate intake, malabsorption or metabolic losses of nutrients due to disease or use of drugs. Use of medication such as oral contraceptives may increase cellular requirements for certain vitamins and minerals because of selective increments in specific binding proteins or enzymes which sequester more of the available nutrient. This may result in maldistribution within the cellular functions. Deficiencies lead to biochemical lesions followed by molecular lesions which lead, often in the presence of other stress factors, to morphological lesions or disease. Except for night blindness due to vitamin A-deficiency, the molecular basis of vitamin deficiency diseases is poorly understood.

The functional impact of marginal malnutrition is still unknown and hence it is impossible to quantify ill health in terms of biochemical lesions which may provide the earliest clue to tissue depletion.

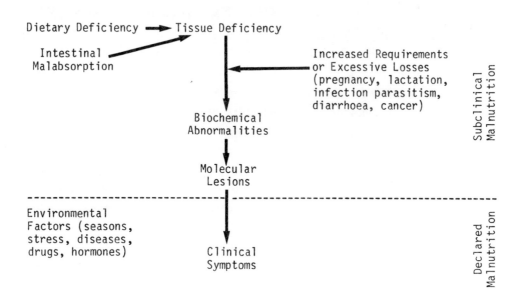

Fig. 1 : Development of nutritional deficiency diseases

INFLUENCE OF CEREAL AND NON-CEREAL FOOD COMPONENTS ON VITAMIN INTAKE

Rice is the staple food consumed in many developing countries. It is now generally accepted that cereal-based diets, if consumed in amounts adequate to meet caloric needs, can fulfill protein needs. However, such diets may be deficient in vitamins and minerals. Recent diet surveys conducted by the National Nutrition Monitoring Bureau (NNMB) of the ICMR in 8 states of India reiterate this point (NNMB Annual Report for 1980). Almost 39% of the households surveyed in rural India had deficient calorie intake. Though 19% had a deficient intake of proteins, this was secondary to a deficiency in calories. Fewer than 1% had a protein deficiency despite an adequate intake in calories (Table I).

On the other hand, the diets were often very deficient in vitamins A and B_2 (Table II). Although average intakes of thiamin, niacin and vitamin C were adequate, diets in some states were deficient in these vitamins too (Table II). Surprisingly, a recent survey from Japan suggests that even in this developed country, the diet is deficient in calories and vitamins, particularly riboflavin (Hosoya, 1977) (Table II).

In India, the type of cereal consumed differs markedly from one state to another. Scrutiny of the data in Table II shows that in states where people eat mixed-cereal diets (Karnataka, Uttar Pradesh and Gujarat) B-vitamin intake is higher than where more than 80% of the cereal eaten is rice. Intake of vitamins A and C was not influenced by the type of cereal, but the data in Table III suggest that the quantity of non-tuberous vegetables and fruits influences the content of these vitamins in the diet.

As expected, intake of all nutrients was influenced by income. Families with per capita daily income of less than Rupees 2/- (US$ 0.25) had inadequate intake of calories as well (Table IV). Income influenced the consumption of food items such as pulses, vegetables, fruits, milk, sugar and oil, but less so for cereals, suggesting that the increased calorie consumption of better-off families came from these relatively high-cost foods (Table V). The deficient intake of vitamins was seen in all income groups. Thus the recent NNMB surveys suggest that the major nutritional

Table I : Percentage distribution of households according to protein (P) and/or calorie (C) adequacy in rural India[a]

State	Sample size	PC --[b]	PC -+	PC +-	PC ++	P -	C -
Kerala	164	31.6	4.3	17.1	47.0	35.9	48.7
Tamil Nadu	400	29.7	0.8	18.3	51.2	30.5	48.0
Karnataka	408	5.6	0.0	7.1	87.3	5.6	12.7
Andhra Pradesh	396	22.2	0.0	15.4	62.4	22.2	37.6
Gujarat	188	12.8	0.0	31.4	55.8	12.8	44.2
Orissa	280	28.6	0.4	12.5	58.5	29.0	41.1
West Bengal	392	14.3	0.8	14.3	70.6	15.1	28.6
Uttar Pradesh	247	3.2	0.0	49.0	47.8	3.2	52.2
Average (Total)	(2475)	18.5	0.8	20.6	60.1	19.3	39.1

(a) Data from National Nutrition Monitoring Bureau, 1981.
(b) - Deficient intake; + Adequate intake.

Table II : Food and nutrient intake in rural India[a] and Japan

| | Calories | Rice as % of | % RDA for 2,400 calories/C unit[b] | | | | | |
	% of RDA	total cereal	Protein	Vit. A	B$_1$	B$_2$	Niacin	Vit. C
Kerala	90	97	100	51	61	64	83	175
Tamil Nadu	90	98	106	31	82	63	86	82
Orissa	103	97	104	61	66	52	95	134
Andhra Pradesh	100	88	103	39	75	59	86	67
West Bengal	107	86	106	61	83	65	108	163
Karnataka	125	65	118	28	152	73	92	33
Uttar Pradesh	103	38	142	28	192	101	164	88
Gujarat	97	27	126	36	160	92	94	71
Average	100	74	113	42	110	70	96	104
Japan[c,d]	91	-	157	81	99	79	-	263

(a) Data from National Nutritional Monitoring Bureau, 1981
(b) C Unit = consumption unit
(c) Data from Hosoya, 1977
(d) % RDA, corrected for 2,400 calories.

Table III : Relationship between intake of vitamins A, B₁, B₂, C and intake of some non-cereal foods[a]

	Pulses	Fruits & vegetables	Milk	Flesh foods	A	B₁	B₂	C
		g/C unit[b]/day				% RDA		
West Bengal	16.2	168.9	52.8	21.3	66	90	65	181
Orissa	32.5	172.1	16.5	11.8	63	68	53	141
Kerala	16.0	186.9	69.5	37.0	47	55	58	167
Andhra Pradesh	29.7	95.1	112.5	13.2	39	75	59	69
Gujarat	37.4	88.7	194.0	–	35	155	88	71
Tamil Nadu	27.9	66.6	75.3	24.9	28	75	58	78
Uttar Pradesh	44.6	91.3	88.6	8.0	28	171	91	82
Karnataka	61.6	38.8	95.3	7.9	28	185	89	42

(a) Data from National Nutritional Monitoring Bureau, 1981
(b) C unit = consumption unit

Table IV : Effect of income on nutrient intake - Andhra Pradesh[a]

Income/Capita/Day	Rupees < 1	1 - 2	2 - 5	≥ 5	All incomes
Number of households	66	152	114	64	376
Protein (g)	52.3	54.8	58.2	63.1	56.7
Fat (g)	19.8	23.2	34.1	50.8	30.2
Calories	2213	2318	2474	2597	2391
Vitamin A (µg)	192	250	291	520	296
B_1 (mg)	0.92	0.84	0.92	0.96	0.90
B_2 (mg)	0.65	0.68	0.81	1.02	0.77
Niacin (mg)	12.3	13.2	13.4	13.9	13.2
Vitamin C (mg)	19.8	29.9	38.3	54.6	34.5

(a) Data from National Nutritional Monitoring Bureau, 1981.

Table V : Effect of income on food intake (g/C unit[a]/day) - Andhra Pradesh[b]

Income/Capita/Day	Rupees < 1	1 - 2	2 - 5	≥ 5	Total
Number of households	66	152	114	64	396
Cereals and millets	541.8	557.5	551.4	500.9	544.0
Pulses	27.4	22.8	32.6	43.4	29.7
Vegetables	31.2	46.7	63.4	86.3	55.3
Roots and vegetables	17.0	17.1	23.4	27.6	20.6
Fruits	15.0	31.9	42.9	78.4	39.8
Flesh foods	9.7	18.5	9.3	11.4	13.2
Milk	33.5	67.3	136.7	258.5	112.5
Fats and oils	11.0	12.5	19.6	29.5	17.0
Sugar and jaggery	5.0	7.7	11.4	17.6	9.9

(a) C unit = consumption unit.
(b) Data from National Nutritional Monitoring Bureau, 1981.

deficiencies in cereal-based Indian diets are vitamins A, riboflavin, other vitamins and calories (in that order). Intake of B-vitamins appears to be influenced by the type of cereal, while that of vitamins A and C is influenced mainly by the quantity of vegetables and fruits consumed.

IMPACT OF B-VITAMIN THERAPY ON RURAL SCHOOLCHILDREN

I would now like to discuss some of our recently published studies on the impact of B-vitamin therapy or low-dose supplements on rural schoolchildren (Bamji et al., 1979; Bamji et al., 1982; Rameshwar Sarma et al., 1981). We became interested in this problem when some of my colleagues observed a very high incidence of angular stomatitis among some rural schoolchildren around Hyderabad. The condition did not respond to treatment with B vitamin complex.

In a systematic investigation, we examined the vitamin status of these children before and after treatment for one month with 2 tablets daily of a multivitamin preparation containing (mg/tablet) thiamin 2, riboflavin 2, pyridoxine 5, calcium pantothenate 2, niacin 20. In one group of boys, the angular lesions were treated with topical application of gentian violet (Bamji et al., 1979). The following biochemical tests were used to assess vitamin status (Sauberlich et al., 1973; Bamji, 1981) :

Vitamin A   Plasma retinol

Thiamin   Erythrocyte transketolase activity and in vitro activation
               with thiamin pyrophosphate (ETK-AC)

Riboflavin   Erythrocyte glutathione reductase activity and in vitro
               activation with FAD (EGR-AC)

Pyridoxine   Erythrocyte aspartate aminotransferase activity and in vitro
               activation with pyridoxal phosphate (EAspAT-AC).

Angular stomatitis (prevalence 41.3%) could neither be treated nor prevented with vitamin therapy (Table VI). However, it improved with the
topical application of gentian violet, suggesting a non-nutritional etiology. Glossitis (prevalence 18.2%) did show improvement and its appearance
could be prevented by treatment with B-vitamin complex (Table VI).

Biochemically, all the boys had marked riboflavin- (as judged by the
glutathione reductase test) and vitamin A-deficiencies. Over 50% also had a
pyridoxine-deficiency, but the prevalence of biochemical thiamin-deficiency
was low (Bamji et al., 1979).

After treatment with therapeutic doses of vitamins for one month, over
50% of the children continued to suffer from riboflavin- and 50% from
pyridoxine-deficiencies using the cut-off points described in the literature
(Sauberlich et al., 1973; Bamji, 1981).

Table VI : Response of oral lesions to B-complex vitamin
therapy for 1 month[a]

	Treatment	Cured	Prevented
		(% of subjects examined)	
Angular stomatitis	B-complex	23	79
	Placebo	12	87
Glossitis	B-complex	94[b]	98
	Placebo	54	90

(a) Data from Bamji et al. (1979)
(b) P < 0.05 compared to placebo

These biochemical results would either indicate that the interpretive guidelines used for enzymatic tests were too rigorous, or that in this community there are non-nutritional constraints which interfere with tissue saturation. The second alternative seems more likely since the scatter in the values was similar before and after treatment. Because of this large scatter, we could not use the data to derive a new set of interpretive guidelines.

IMPACT OF LOW-DOSE, LONG-TERM VITAMIN SUPPLEMENTS IN RURAL SCHOOLBOYS

In the same community, a second study was carried out with the following objectives (Rameshwar Sarma et al., 1981; Bamji et al., 1981) :

1. Assess the seasonal variations in the clinical signs attributable to B-vitamin deficiency and minor ailments.

2. Examine the impact of long-term low-dose (RDA) vitamin supplements on :

   a) the prevalence of vitamin deficiency signs,

   b) the biochemical status of vitamins B_1, B_2, B_6 and folic acid,

   c) selected parameters of psychomotor performance, and

   d) growth and minor ailments.

3. Derive discriminatory guidelines for the interpretation of enzymatic tests, hoping that tissue saturation would be achieved after prolonged supplementation.

Three hundred and ninety-seven schoolboys aged 6-13 years, were either given a tablet of B-complex vitamins containing (mg/tablet) thiamin 1, riboflavin 1.5, pyridoxine 1.5, niacin 15, folic acid 0.1 and vitamin B_{12} 0.001, or a similar-looking placebo, over a period of 1 year. Both kinds of tablets were specially prepared by MM. Biological Evans - Hyderabad, India. Clinical and anthropometric examinations were carried out at the beginning of the study and at 4 month intervals. History of minor ailments such as gastrointestinal disorders and respiratory infections was taken at the time of each examination. As a marker of tablet consumption, urinary excretion of riboflavin was measured in random samples of urine, collected from 15 subjects in each group at every time point. Biochemical parameters were measured after 12 months of treatment, in a subsample of sixty boys who had consumed the tablets for at least 80% of the period. Psychomotor performance was also tested in the same children.

Seasonal variations in the incidence of nutritional deficiency signs, infections and urinary excretion of riboflavin

The highest incidence of respiratory infections was observed in winter (November) whereas gastrointestinal disorders tended to occur more often during the monsoon season (July). In both groups, urinary riboflavin excretion tended to be highest in winter (Table VII). Whether this is causally related to the high rate of infection in that season needs to be investigated. Prevalence of glossitis tended to increase with the onset of summer (March) (Table VIII).

Effect of B-vitamin supplements on growth, point prevalence of vitamin deficiency signs and morbidity

Vitamin supplements had no effect on growth or point prevalence of respiratory or gastrointestinal ailments (Table VII). Even the prevalence of vitamin-deficiency signs was similar in the treated and placebo groups, suggesting that the low-dose supplements did not have any clinical impact (Table VIII).

Impact of B-vitamin supplements on biochemical parameters

The vitamin status of the supplemented children was significantly superior to the placebo group as judged by the biochemical tests (Table IX). However, after vitamin intakes at levels higher than RDA for one year, over 50% of the children were still deficient (particularly in riboflavin). The reasons for this are not clear. It is possible that frequent bouts of infection resulted in excessive urinary losses of vitamins, preventing the tissues from reaching saturation. The rise in the incidence of glossitis in March may be a delayed effect of excessive tissue depletion during winter.

Table VII : Point prevalence of minor ailments and trends in urinary riboflavin in rural schoolboys at different periods (data from Rameshwar Sarma et al., 1981)

	Groups	Period of study			
		March '79	July '79	November '79	March '80
				Point prevalence	
Respiratory Tract Infections	Placebo	9.6	9.6	38.3	30.6
	Supplemented		9.9	41.5	31.6
	Combined		9.7[a]	39.9[b]	31.1[c]
Urinary riboflavin	Placebo	90.0 ± 11.8[a]	268.7 ± 59.4[b]	641.4 ± 214.4[a]	191.2 ± 27.1[b]
		(14)	(20)	(17)	(17)
	Supplemented	123.5 ± 26.2[a]	890.6 ± 114.2***[b]	1494.6 ± 239.3***[c]	581.3 ± 113.4***[b]
		(18)	(19)	(20)	(17)

*** $P < 0.01$ compared to placebo group.

Figures in parentheses indicate the subsample size.

Values in a row bearing different superscripts are significantly different ($P < 0.05$).

Normal proportions t-test, independent t-test (two-tailed) and modified t-test applied wherever appropriate.

Table VIII : Point prevalence of clinical signs in rural schoolboys at different periods
(Data from Rameshwar Sarma et al., 1981)

Period of study	Angular stomatitis		Glossitis		Bitot's spots	
	PL	SUP	PL	SUP	PL	SUP
March'79	24.7	25.6	18.7[a]	19.6[a]	11.6	12.6[a]
July'79	25.9	24.0	8.5[b]	18.4*[a]	8.5	12.3[a]
November'79	24.7	21.6	7.7[b]	7.0[b]	6.6	5.8[b]
March'80	21.1	25.1	10.3[b]	12.9[a]	8.6	4.7[b]

PL   Placebo
SUP   Supplemented group
a,b   In a column values bearing different superscripts are significantly different by two-tailed proportion test.
*   P < 0.05 compared to placebo group by two-tailed proportion test.

Table IX : Percentage distribution of boys according to biochemical status, with mean values[a]

Treatment	Mean ± S.E.	Percent distribution		
		Low risk	Medium risk	High risk
EGR – AC (Riboflavin)		< 1.2	1.2 – 1.4	≥ 1.4
Placebo[b]	1.83 ± 0.04	0	5.2	94.8
Supplement	1.43 ± 0.03***	17.9	32.1	50.0
EASPAT – AC (Pyridoxine)		< 1.7	1.7 – 2.00	≥ 2.0
Placebo	1.86 ± 0.03	29.3	46.6	24.1
Supplement	1.66 ± 0.03***	55.4	41.1	3.5
RBC – Folate		≥ 160	140 – 160	< 140
Placebo	133.9 ± 5.6	26.4	18.9	54.7
Supplement	187.8 ± 7.9***	65.1	14.0	20.9

(a) Data from Bamji et al., 1982

(b) Number of subjects : placebo, 58; supplemented, 56;

*** P < 0.001, compared to placebo by Student's t test (two-tailed).

Impact of vitamin supplements on psychomotor performance

Of the 4 tests, including cubes intelligence test (mental age) and three psychomotor tests (reaction time, finger dexterity, and hand steadiness), the supplemented children performed significantly better in the steadiness test (Table X). Riboflavin-deficient children tended to make more errors (Table XI), suggesting a correlation between hand steadiness and riboflavin status.

On the basis of these data, we propose that, in a community where the prevalence of infections is high, there may be periodic metabolic losses of vitamins in the urine. This may interfere with tissue saturation despite adequate intake. Dietary improvement in such a community may not achieve the expected clinical impact in terms of reduced incidence of deficiency signs and symptoms. Seasonal variations in the incidence of nutritional deficiency signs may be more closely linked to morbidity than to intake of dietary vitamins. Even in the absence of clinical signs, some improvement in bio-chemical status and selected psychomotor performance tests may be achieved through vitamin supplements. The high incidence of angular stomatitis in this community appears to be unrelated to B-vitamin deficiency.

In a recent study in Gambia, Bates et al. (1981) found that administra-tion of food supplements which raised riboflavin intake to levels close to RDA, did not normalise the EGR-AC in rural pregnant women. The authors feel that the vitamin requirement of some rural communities may be higher. In the case of the Hyderabad study, one can argue that in the absence of an adequate supply of calories and proteins, vitamins may not have been utilised. The same argument however would not apply to the Gambian study where food supplements were given.

The important questions are how much higher is the vitamin requirement of such communities and what should be the practical strategy to meet it ?

Table X : Psychomotor performance of B-complex vitamins supplemented
and unsupplemented boys[a]

Test		Treatment	
		Placebo	Supplement
Mental age	(years)	9.8 ± 0.44[b]	10.9 ± 0.49
Steadiness test	(error scores)	332.1 ± 17.3[***]	246.0 ± 13.7
Reaction time	(seconds)	702.0 ± 18.8	706.7 ± 18.4
Finger dexterity test	(seconds)	525.4 ± 25.9	500.0 ± 18.7

(a) Data from Bamji et al., 1982
(b) Mean ± S.E.M.
*** P < 0.001 compared to supplemented group by modified t-test
     (two-tailed).

Table XI : Relationship between riboflavin status and errors committed
in steadiness test - per cent distribution[a]

Group[b]	Errors committed	EGR - AC	
		< 1.4	1.4
1	< 170	39.1	60.9
2	170 - 420	28.2	71.8
3	> 420	9.5[*]	90.5[*]

(a) Data from Bamji et al., 1982
(b) Values for treatment groups were pooled.
*   Significantly different from group 1 by proportion test for small
     samples, P < 0.05.

ACKNOWLEGEMENTS

The author is grateful to Dr. P.G. Tulpule, Director, National Institute of Nutrition, Dr. S.G. Srikantia, Ex-Director, National Institute of Nutrition and several other colleagues including Dr. M.C. Swaminathan, Dr. Prahlad Rao and Shri Nadamuni Naidu, for useful discussions.

REFERENCES

Bamji, M.S. (1981) Laboratory tests for the assessment of vitamin nutrition status. In : Vitamins in Human Biology and Medicine (Ed. Briggs, M.) CRC Press Inc., Boca Raton, Florida, pp. 1-27.

Bamji, M.S., Rameshwar Sarma, K.V. & Radhaiah, G. (1979) Relationship between biochemical and clinical indices of B-vitamin deficiency. A study in rural school boys. Brit. J. Nutr. 41:431-441.

Bamji, M.S., Arya, S., Rameshwar Sarma, K.V. & Radhaiah, G. (1982) Impact of long-term, low-dose vitamin supplements on vitamin status and psychomotor performance of rural school boys. Nutr. Res. 2:147-154.

Bates, C.J., Prentice, A.M., Paul, A.A., Sutcliffe, B.A., Watkinson, M. & Whitehead, R.G. (1981) Riboflavin status in Gambian pregnant and lactating women and its implications for Recommended Dietary Allowances. Am. J. Clin. Nutr. 34:928-935.

Hosoya, N. (1977) Health Aspects of Community Development in South East Asia (Ed. Katsunuma, H. & Maruchi, N.).

National Nutrition Monitoring Bureau (1981) Annual Report for the year 1980, National Institute of Nutrition, Hyderabad, India.

Rameshwar Sarma, K.V., Radhaiah, G. & Bamji, M.S. (1981) Impact of long-term, low-dose B-complex vitamin supplements on clinical and anthropometric status of rural school boys. Nutr. Rep. Int. 24:345-353.

Sauberlich, H., Dowdy, R.P. & Skala, J.M. (1973) Laboratory tests for assessment of nutritional status. Crit. Rev. Clin. Lab. Sci. 4:215-340.

VITAMIN A-DEFICIENCY IMPAIRS THE NORMAL MANNOSYLATION, CONFORMATION AND IODINATION OF THYROGLOBULIN : A NEW ETIOLOGICAL APPROACH TO ENDEMIC GOITRE

Yves INGENBLEEK

Nestlé Products Technical Assistance Co. Ltd, Research Department, CH-1814 La Tour-de-Peilz, Switzerland

SUMMARY

This study was undertaken in order to validate the hypothesis that vitamin A-deficiency alters the structure of thyroglobulin (Tg). For that purpose, four groups of 20 Sprague-Dawley rats were submitted during two months to varying dietary conditions, namely a control diet (C^+), a vitamin A-deficient diet (A^-), an iodine-deficient diet (I^-) and a diet characterized by the association of both deficiencies (A^-I^-). Both the conventional parameters of thyroid function, the intracellular steps of Tg glycosylation and iodination were analyzed. In the A^- and A^-I^- groups, blood levels of retinol fell to one tenth of the control mean and circulating concentrations of total and free T_4 and T_3 increased significantly. This biochemical hyperthyroidism contrasted with the maintenance of normal TSH plasma values, suggesting a generalized peripheral refractoriness to thyroid hormones. In both A^- and A^-I^- groups, thyroid cytosol ^{3}H-RPM (retinyl-phosphate-mannose) and ^{3}H-mannose incorporation into the core of the 12S-Tg and 19S-Tg species were reduced by 40-50%. In contrast, cytosolic concentrations of ^{3}H-DPM (dolichyl-phosphate-mannose) rose, suggesting that the N-glycosylation pathways are affected in opposite direction. The sedimentation coefficient in sucrose gradient of the purified dimeric ^{125}I-19S-Tg after guanidine 6M and dithiothreitol denaturation showed that most of the A$^-$ Tg molecules were transformed into monomeric 12S species, implying alterations of both noncovalent and covalent bonds. Finally, the radiochromatogram of ^{125}I-iodothyronines recovered after Tg pronase digestion revealed a significant increase in the mono- (MIT) and diiodothyronine (DIT) fractions in contrast with a significant decrease in the T_3 and T_4 hormonal compounds. These findings are consistent with the view that vitamin A-depletion impairs the endogenous RPM synthesis and, therefore, the normal Tg O-mannosylation. The growing peptide is characterized by steric hindrance, leading to abnormal closure of disulphide bonds, reduced MIT-DIT coupling reactions and depressed generation of physiologically active thyroid hormones. Pure iodine deficit (I$^-$) induces no effects on the above-mentioned glycosylation reactions, but iodine shortage superimposed on preexisting vitamin A-deficit (A$^-$I$^-$) aggravates the Tg dysmaturation. This experimental investigation confirms at all points the molecular mechanism of a vitamin A-dependent goitrogenic process predicted on the basis of epidemiological data recorded in Senegalese patients (Y. Ingenbleek & M. De Visscher, 1979).

* * *

The first descriptions of goitrous hyperplasia were first mentioned in ancient Chinese, Indian and Egyptian documents. Before the turn of the present century, many speculations had been made as to the possible etiological factors responsible for its morbid development and, correspondingly, numerous treatments were proposed (Kelly & Snedden, 1960).

Around 1850, low iodine levels in drinking water were regarded as the main cause of goitre (Maffoni, 1846). Later, low iodine content of the atmosphere, soils and natural foodstuffs were involved as contributory factors (Chatin, 1853). The discovery of unusually high iodine content in the thyroid gland (Baumann, 1895) further corroborated the iodine deficiency theory. In 1920, following a proposal by Marine and Kimball, iodized salts were used in areas where goitre was prevalent. More recently, this therapeutic approach was extended by the widespread use of iodized oil (McCullagh 1963). These public health prophylactic measures had considerable success in reducing the incidence of endemic goitre and confirmed beyond doubt that the halide shortage played a central role in the etiopathogenesis of the disease.

Nevertheless, the probable coexistence of other etiological factors became increasingly clear and was already perceived six decades ago in the following statement : "The iodine deficiency theory is attractive at first sight, but there are many matters which cannot be explained by this theory alone" (Anonymous, 1924). Despite continuing advances obtained in all aspects of iodine metabolism and thyroid function both in normal and goitrous subjects, no substantial gain in knowledge was achieved regarding associated etiological factors. In order to fill the gap between the iodine deficit theory and epidemiological data, several plausible explanations, including genetic factors, naturally occurring goitrogens of vegetal origin, content of drinking waters in minerals and trace elements, bacterial pollution, and generalized malnutrition, were tentatively evoked (Koutras et al., 1973, Delange, 1974; Gaitan, 1980; Koutras, 1980; Medeiros-Neto, 1980). This problematic situation was summarized by J.B. Stanbury who pointed out that "it is necessary to look for other environmental factors which, working in concert with iodine deficiency or even alone, may ensure the persistence and even provoke the emergence of endemic goitre" (1969). The same author emphasizes that "other factors, both dietary, genetic and possibly immunologic, may be responsible for the substratum of thyroid disease which remains in a population group which is receiving an abundance of iodine in the diet" (1973).

On the basis of epidemiological data recorded in goitrous patients living in the southern part of Senegal, Ingenbleek and De Visscher have recently observed that the thyroid swelling was negatively correlated with the patients' retinol status (1979). Inducing from a study performed on the rat

α_1-macroglobulin (Kiorpes et al., 1976), they postulated that the deleterious effects of poor nutritional status on thyroid function could be mediated through the defective formation of the RPM (Fig. 1) precursor necessary for the normal Tg mannosylation and maturation.

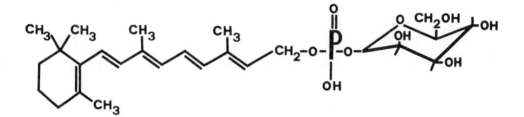

Fig. 1 : Retinyl-phosphate-mannose (RPM)

The proposed explanation was an impaired incorporation of mannose in the growing peptide, resulting in failure to close the disulphide bonds. This view assumed that the abnormal spatial conformation of the Tg glycoprotein chains hampered the adequate coupling reactions between iodotyrosine residues, leading to the appearance of an iodine-independent goitrogenic process. The present experiments were undertaken in order to validate the hypothesis. The results unequivocally endorse the concept.

MATERIAL AND METHODS

Eighty weanling male Sprague-Dawley rats weighing approximately 55 g were obtained from Simonsen laboratories, Inc., Gilroy, California, and were assigned to four groups of twenty rats, each of them designated by a classical symbol (Coplan & Sampson, 1935).

Dietetic management

The first group (C^+) was fed <u>ad libitum</u> a control semi-synthetic Purina lab chow diet (Table I).

Table I : Composition of experimental diets

Nutriment	Content	Animal groups			
		C^+	A^-	I^-	A^-I^-
Cornstarch	37%	+	+	+	+
Glucose, anhydrous	29%	+	+	+	+
Vitamin-free casein	18%	+	+	+	+
Cottonseed oil	5%	+	+	+	+
Salt mix[a]	4.7%	+	+	+	+
Vitamin premix[b]	3.6%	+	+	+	+
Cellulose	2.7%	+	+	+	+
Vitamin A (acetate)	9,900 IU/kg	+	−	+	−
I^- (potassium iodide)	170 μg/kg	+	+	−	−

(a) Williams-Briggs salt mixture without iodine, obtained from Nutritional Biochemical Company, Cleveland, Ohio.

(b) Devoid of vitamin A. Each kg of the dry diet provided ergocalciferol, 1,800 IU; α-tocopherol, 110 mg; phytylmenaquinone, 1 mg; ascorbic acid, 1 g; inositol, 110 mg; choline chloride, 1.65 g; p-amino-benzoic acid, 110 mg; niacin, 100 mg; riboflavin, 22 mg; pyridoxine-HCl, 22 mg; thiamin-HCl, 22 mg; calcium pantothenate, 66 mg; folic acid, 2 mg, biotin, 440 μg; cyanocobalamin, 29.7 μg.

The C^+ diet was supplemented with 3,400 μg (or 9,900 IU) of vitamin A acetate (Hoffmann-La Roche) and 170 μg iodide/kg diet in the form of KI. This C^+ diet is believed to fulfill the optimal requirements for normal growth.

The second group (A⁻) was fed with the stock diet. enriched with only 170 µg iodide/kg. The third group (I⁻) is supplemented with the stock diet containing only the vitamin A load. The fourth group (A⁻I⁻) was fed the stock diet without any additions, and was, therefore, deficient in both vitamin A and iodine.

Experimental protocol

The rats were individually housed in cages, in an air-conditioned and light-controlled room with a mean temperature of 21°C, and are weighed every three days. All rats had free access to food and drinking water. The daily food intake of each animal was evaluated. During the experiment, the food intake of the deficient rats was similar to that of the normal group. After two months, all animals were killed by ether anaesthesia and then exsangui-nated. Their thyroid glands were removed and plasma frozen at -60°C until analyzed for retinol. In each group, 13 rats were devoted to the study of the blood indicators of the thyroid function and to the intrathyroidal steps of the glycosylation reactions. The 7 remaining rats were injected by intra-peritoneal route, 24 h prior to sacrifice, with 25 µCi carrier-free ^{125}I-Na (Amersham, U.K.) in sterile and isotonic (NaCl 0.9 g/dl) solution. This permitted the further study of the purified Tg profile and the radio-chromatogram of the Tg-bound iodothyronines.

Blood parameters

Retinol was analyzed fluorometrically in the isopropanol eluate by use of an Amino-Bowmann spectrofluorometer according to Garry et al., 1970. Fresh solutions of all-trans-retinyl acetate are used as a standard after determination of their absorption at 328 nm ($E_{1cm}^{1\%}$ = 1835 in hexane).

Both total thyroxine (TT_4) and total triiodothyronine (TT_3) were measured by RIA with Amersham kits, UK. Free thyroxine (FT_4) and free triiodothyronine (FT_3) levels were determined by the double labelling Sephadex-binding technique (Finucane et al., 1976). Samples were assayed in duplicate and simultaneously counted for both ^{131}I and ^{125}I content,

employing a multichannel Packard Gamma-Counter. Results were corrected for
[131]I spillover into the ^{125}I channel and for machine efficiency. FT_4
and FT_3 were calculated by multiplying total hormonal level by the
dialyzed free hormonal fraction. Reverse triiodothyronine (rT_3) analysis
was performed with Serono RIA kits. TSH was measured in duplicate by a
double antibody RIA using a NIAMDD kit (supplied by NIH Rat Pituitary
Hormone Distribution Program) and expressed as ng TSH/ml of NIAMDD TSH-RP-1
standard.

Preparation of the ^{3}H-labelled subcellular organelles

After weighing, the thyroid glands of 13 rats in each group were
incubated in 3.5 ml of Krebbs-Ringer phosphate (KRP) buffer medium[c]
containing 0.9 mCi ^{3}H-mannose of high specific activity (28 mCi/mg,
Amersham, U.K.). Incubation procedure was performed under dark conditions at
pH 7.4, and lasted 120 min. at 37°C. After 2 hr, the thyroid glands were
washed and rinsed at room temperature in 4 ml of the KRP buffer medium. The
thyroid tissue was minced with scissors and homogenized in 7 volumes of TKM
buffer medium[d] using a loosely fitting Elvehjem homogenizer. The resulting
homogenate was submitted to differential centrifugation at 2,500 g for
20 minutes, and the supernatant at 105,000 g (29,000 rpm) for 60 minutes in
a SW 65 rotor of an International Ultracentrifuge. This last supernatant
contained the soluble proteins (among which are stable Tg and ^{3}H-Tg) and
was stored at -23°C until further assay. The pellet, comprising mitochondria, lysosomes, and microsomes, was resuspended in a small volume of
saline and stored in liquid N_2 until analyzed. An aliquot of both
supernatant and pellet fractions was utilized for the simultaneous
estimation of the residual radioactivity and protein content (Lowry et al.,
1951) with bovine serum albumin as a standard. Recovery was 95% and 80%,
respectively.

(c) KRP : 143 mM NaCl + 5.6 mM KCl + 1.42 mM $MgSO_4$ + 3 mM $CaCl_2$ + 10 mM
    sodium phosphate buffer, pH 7.4.

(d) TKM : 0.35 M sucrose containing 25 M KCl, 5 mM $MgCl_2$, 50 mM Tris-HCl
    buffer, pH 7.6.

Measurement of the mannolipid fractions

Extraction of RPM and DPM from thyroid pellet fractions followed the same methods described for other animal organs (Silverman-Jones et al., 1976; De Luca, 1977; Masushige et al., 1978). All steps were conducted in red-lighted room. Separation of the two mannolipids by differential solvent extraction utilizes the RPM property to be more hydrophilic than DMP. 5 volumes (1 ml) of chloroform/methanol (2:1, v/v) was added to the pellet mixture. The tube was stirred and allowed to separate by low-speed centrifugation. Under these extraction conditions, approximately 98% of DPM was recovered in the lower phase, whereas RPM partitions about 40% in the lower phase and 60% in the upper phase. Then, 15 vol. (3 ml) of chloroform/methanol (2:1, v/v) was added to yield a monophasic extract which was dried under a stream of N_2, dissolved in 2 ml of 99% methanol, and applied to a column (1 x 5 cm) of DEAE-cellulose-acetate. The first elution with 250 ml of 99% methanol removed all impurities, and the latter one with 25 ml of 0.050 M ammonium acetate in 99% methanol released both mannolipids. This eluate was dried under N_2 conditions and redissolved in 2.5 ml of chloroform/methanol (2:1, v/v). Purity of RPM and DPM was assessed by silicagel thin-layer chromatography with a chloroform : methanol : water (60:25:4) system according to Tkacz et al. (1974), allowing a consistent discrimination between RPM (Rf 0.25) and DPM (Rf 0.50). The dry plates were scraped and sections (0.5 cm) of silicagel were collected into counting vials. Radioactivity was determined after addition of 0.25 ml of methanol and 10 ml of Betafluor (National Diagnostics, Somerville, NJ, USA).

Isolation of labelled ^{3}H-12S-Tg and ^{3}H-19S-Tg in the thyroid cytosol

This approach allowed the quantitative evaluation of ^{3}H-mannose incorporation into the core of the Tg peptides. In each of the four investigated groups, the soluble proteins were pooled and precipitated by the addition of $(NH_4)_2SO_4$ to 48% saturation at pH 6.8. After centrifugation, the pellet was dissolved in 2 ml of PBS buffered saline[e] and the proteins were again precipitated with $(NH_4)_2SO_4$. After dialysis against PBS overnight

---

(e) PBS : 10 mM sodium phosphate buffer, 150 mM NaCl, pH 6.8

at 4°C, the solution was centrifuged in a SW 65 rotor of an International Ultracentrifuge at 29,000 rpm (105,000 g) for 60 minutes to remove insoluble materials. Sucrose density linear gradient (5-20% w/v) was performed in 11 ml of PBS at 105,000 g for 15 h at 4°C. After centrifugation, 50-52 fractions were collected from each tube with a peristaltic pump, the first fraction starting from the bottom. In each tube, the protein content was read by absorbance after suitable dilution, assuming that the $E_{280}^{1\%}$ for thyroglobulin is 10.0 (Sorimachi & Ui, 1974). Radioactivity of ^{3}H-Tg was measured using 0.1 ml of the solution. Results obtained both for the protein content and β-counting were normalized, implying that they are expressed as a percentage of the maximal value.

Isolation of labelled ^{125}I-12S-Tg and ^{125}I-19S-Tg in the thyroid cytosol

This approach was necessary for the determination of the Tg radioiodinated profile. In each of the four investigated groups, the soluble proteins were pooled and submitted to the same procedure as the one described after ^{3}H-mannose incubation. However, and in order to remove free radioiodine completely, two additional precipitation steps with $(NH_4)_2SO_4$ were performed. Final precipitates were dissolved in 2 ml of PBS. ^{125}I was counted in a Packard Auto-Gamma scintillation counter. Results obtained for γ-counting were normalized.

Study of the Tg noncovalent bonds

The supernatant containing soluble proteins and ^{125}I-Tg was layered on a Sepharose 6B column. Elution was performed with 0.05 M phosphate buffer at pH 7. Purified Tg was dissolved in 6 M guanidine-HCl in 0.1 M sodium phosphate-medium at pH 7.2 for 4 h at 20°C. The ^{125}I-Tg molecules were then partially denatured by the reduction of most noncovalent bonds. The sample was submitted to sucrose density linear gradient (5-15% w/v) in 6M guanidine-HCl pH 7.0 for 20 h at 49,000 rpm (230,000 g). Fractions of five drops each were collected and allotted to scintillation in a Packard Auto-Gamma well-counter. Sedimentation coefficients of the individual peaks were determined by comparison with the location of the major ^{125}I-Tg peak which was assumed to have a sedimentation coefficient of 19S. The proportion of each 12S-Tg and 19S-Tg peak was calculated by integrating the areas under the peaks. Results obtained were normalized.

Study of the Tg covalent bonds

This approach has required a maximal denaturation of the ^{125}I-Tg molecule and the cleavage of most disulphide bonds. For that purpose, pretreated Tg molecules incubated in guanidine-HCl medium, as previously indicated, were resubmitted to the 0.1 M sodium phosphate medium at pH 7.2 in the presence of a molar excess of 100 moles dithiothreitol (DTT) per protein disulphide bond. This procedure was followed by the same sucrose density gradient study for 20 h, γ-counting, and integration operations as mentioned above. Results obtained were normalized.

Determination of ^{127}I-19S-Tg and ^{125}I-19S-Tg radiothyronine profile

Stable iodine (^{127}I) was measured with a Technicon Auto-Analyzer according to the method of Benotti et al. (1965). Analysis of the radio-iodinated Tg profile necessitated a prerequisite pronase digestion. For that purpose, 1 mg ^{125}I-19S-Tg was dissolved in 1 ml of 0.1 M Tris-HCl buffer, pH 8.0. Digestion is performed by the addition of 0.1 mg of pronase (Calbiochem) and 12.5 μmol of methylmercaptoimidazole at 37°C for 20 h in the dark and under N_2 conditions. Iodoaminoacids were analyzed by the method of Osborn & Simpson (1968). A fraction of the sample (0.5 ml) was transferred to the top of a Sephadex G-25 fine column (1.3 x 15 cm) which was previously equilibrated with a mixture of pyridine, acetic acid, and water (45:11.5:1,943.5) (v/v/v) at pH 5.8, and eluted with the same solution. When MIT elution was achieved, the medium was replaced with 0.05 N NaOH to elute the next DIT, T_3 and T_4 peaks. The proportion of each labelled compound was determined with a digital integrator as a percentage of total radioactivity.

Data analysis

All data were evaluated by analysis of variance (Adler & Roessler, 1968). P values of 0.05 or less were regarded as significant. Means were subjected to Duncan's multiple range test when F ratios were significant with one-way analysis of variance. When significant F ratios were found by two-way analysis of variance, overall treatment means were tested for significant differences by the least significant differences test. Pooled standard deviation of the individual treatment means was calculated from the error variance for each variable.

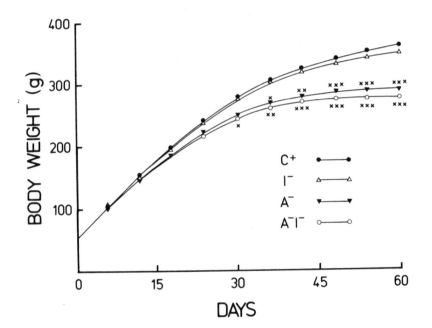

* $p < 0.05$
** $p < 0.01$
*** $p < 0.001$

Fig. 2 : Mean weight curves obtained in the four experimental groups.

RESULTS

The mean body weight profiles of the 4 groups of 20 rats are shown in Fig. 2. The two C^+ and I^- groups demonstrated a comparable evolution, reaching after two months 343 ± 28 g and 339 ± 31 (S.D.) g, respectively. The two A^- and A^-I^- groups had significant growth retardation at the end of the 6th week and this trend was more pronounced for the A^-I^- group. At the time of their sacrifice, the mean weight was 286 ± 17 g $(0.001 < p < 0.01)$ and 273 ± 19 g $(p < 0.001)$, respectively. Both A^- and A^-I^- were typically characterized by ocular inflammatory lesions.

The mean thyroid weight and the circulating parameters of the retinol status and thyroid function on day 60 are collected in Table II, revealing that thyroid weight was not affected by pure vitamin A-deficiency, but fourfold increased by iodine deprivation.

Blood retinol levels were one tenth that of the control $(p < 0.001)$ in both A^- and A^-I^- groups, but were unchanged in the I^- group. TT_4, FT_4, TT_3 and FT_3 were all significantly increased in both the A^- and A^-I^- groups. In the I^- group, the decrease in blood TT_4 and FT_4 was accompanied by increase in plasma TT_3 and FT_3 levels. rT_3 concentrations were significantly increased in the A^- group and significantly decreased in the I^- group, in contrast with the baseline pattern maintained during the entire experiment in the A^-I^- group. The pituitary function, appraised by the circulating TSH levels, was not changed in the A^- group, but was significantly hyperactive in the I^- group. In the A^-I^- group, TSH secretion was intermediate between the A^- and I^- groups.

Table III shows that ^{3}H-mannose partitioned between the two mannolipid fractions. The data indicate that both A^- and A^-I^- groups are characterized by a significant depression in endogenous synthesis of the retinol-linked sugar amounting to 50-60% of control, and a slight increase (10-15%) in monosaccharide uptake by the dolichol systems.

Table II : Thyroid weight, retinol status and circulating parameters of the rat thyroid function (mean ± S.D.)

Rat groups (n = 13)	Thyroid weight (mg)	Retinol µg/dl	TT_4 µg/dl	FT_4 ng/dl	TT_3 ng/dl	FT_3 pg/dl	rT_3 ng/dl	TSH ng/ml
C^+	19.2 ± 2.8	52.1 ± 4.8	5.9 ± 0.9	3.2 ± 0.6	73 ± 11	220 ± 31	405 ± 82	248 ± 32
A^-	15.7 ± 2.6 N.S.	4.9 ± 0.6 ***	9.4 ± 1.1 **	5.4 ± 0.9 **	128 ± 23 ***	418 ± 63 ***	608 ± 107 **	227 ± 29 N.S.
I^-	85.2 ± 15.4 ***	51.7 ± 4.9 N.S.	3.6 ± 0.7 **	1.9 ± 0.4 **	167 ± 39 ***	552 ± 74 ***	225 ± 41 **	915 ± 103 ***
A^-I^-	31.9 ± 6.9 **	4.6 ± 0.5 ***	6.8 ± 0.9 *	4.7 ± 1.2 **	131 ± 29 ***	375 ± 36 ***	396 ± 53 N.S.	381 ± 53 **

N.S. not significant
* $0.01 < p < 0.05$
** $0.001 < p < 0.01$
*** $p < 0.001$

Table III : Subcellular distribution of ^{3}H-mannose

| Rat groups | TTRa | TMRa | RPM | | DPM | |
(n = 13)	10^6 dpm		dpm	% TMRa	dpm	% TMRa
C$^+$	8.91	1.04	1,120	0.107	12,688	1.22
A$^-$	17.04	2.69	1,673	0.062	37,005	1.37
I$^-$	12.66	1.77	1,945	0.109	20,709	1.17
A$^-$I$^-$	18.83	2.46	1,307	0.053	34,895	1.41

Total radioactivity is expressed as number of disintegrations per minute (dpm). Each value is the mean of two measurements made on pooled fractions. Total thyroid radioactivity (TTRa) was measured after incubation and homogenization. Total microsomal radioactivity (TMRa) usually comprised 10 to 20% of TTRa. Recovery was above 90%.

Table IV displays the distribution of ^{3}H-mannose between the 12S-Tg and the 19S-Tg peaks. The normalized results demonstrated that both A$^-$ and A$^-$I$^-$ Tg molecules were characterized by a significantly depressed incorporation of the labelled material ranging from 53% to 64% of the control radioactivity/absorbance ratio.

Data obtained for the analysis of the Tg noncovalent bonds after 6M guanidine incubation are collected in Table V. Under this denaturing procedure, the C$^+$ group was characterized by a major 19S-Tg peak representing 64% of the total radioactivity, whereas the 12S-Tg peak bound the remaining 36%. In both A$^-$ and I$^-$ groups, the relative proportion of the monomeric 12S-Tg peak increased at the expense of the dimeric 19S Tg peak. Maximal denaturation of the mature 19S Tg species was reached in the experimental group combining both A$^-$ and I$^-$ deficiencies. In this group, four fifths of the totally bound ^{125}I radioactivity was taken up by the monomeric 12S-Tg species. Fig. 3 outlines the relative proportion of the two main Tg molecular species after guanidine treatment in each of the 4 experimental groups.

Table IV : Incorporation of ^{3}H-mannose into 12S-Tg and 19S-Tg molecules

Rat groups (n = 13)	Monomeric 12S-Tg				Dimeric 19S-Tg			
	% ^{3}H	% 210 nm	$\frac{Ra}{Ab}$	Ratio	% ^{3}H	% 210 nm	$\frac{Ra}{Ab}$	Ratio
C$^+$	23.7	5.3	4.47	1.00	31.6	33.0	0.95	1.00
A$^-$	16.2	6.1	2.66	0.59	25.4	41.3	0.61	0.64
I$^-$	22.3	5.4	4.13	0.92	29.6	31.7	0.93	0.97
A$^-$I$^-$	12.3	5.2	2.37	0.53	21.8	39.8	0.54	0.56

Each value is the mean of two measurements made on pooled fractions.

Values calculated for ^{3}H-Radioactivity (Ra) and 210 nm Absorbance (Ab) are given as percentages of the profile obtained after sucrose density gradient centrifugation and 2 h ^{3}H-mannose incubation. Data are normalized, implying that the Ra/Ab ratio found for both 12 S-Tg and 19 S-Tg in the C$^+$ group are shifted towards an arbitrary 100% value, and that the same Ra/Ab recorded in both 3 other experimental groups is plotted against this 1.00 reference value.

Table V : Chromatographic study of the Tg noncovalent bonds

Rat groups (n = 7)	Iodine atoms/ mole Tg	Monomeric ^{125}I-12S Tg	Dimeric ^{125}I-19S Tg
C^+	35	36	64
A^-	34	59	41
I^-	16	47	53
A^-I^-	13	81	19

Each value is the mean of two measurements made on pooled fractions. Value obtained for each Tg peak is the percentage of the totally bound radioactivity.

The study of the Tg covalent bonds was performed after DTT incubation of the guanidine-pretreated Tg molecules. Under the effect of this double denaturating procedure, all 19S-Tg molecules were cleaved into their 12S-Tg subunits. The very long subsequent ultracentrifugation, in the absence of DTT, allowed a partial reoxidizing of some cystein residues and partial recovery of disulphide bonds. Table VI and Fig. 4 showed that this is the case for both the C^+ and I^- groups. In contrast, no reaggregation of the mature 19S-Tg species was recorded in either the A^- or in the A^-I^- groups. An additional minor slow-migrating 3-8 S compound was observed in all derivations.

Table VII revealed the profile of the radiothyronines released by purified ^{125}I-Tg after pronase digestion. The free radioiodide represented less than 2% of the total retrieved radioactivity. This Table shows that the I^- group exhibited the expected increase of labelled MIT and T_3, contrasting with the decrease of DIT and T_4, and evolutionary pattern leading to high MIT/DIT and T_3/T_4 ratios. The A^- group showed that both MIT and DIT fractions were significantly elevated, whereas the T_3 and T_4 fractions were reduced by a half and by one fourth, respectively. The

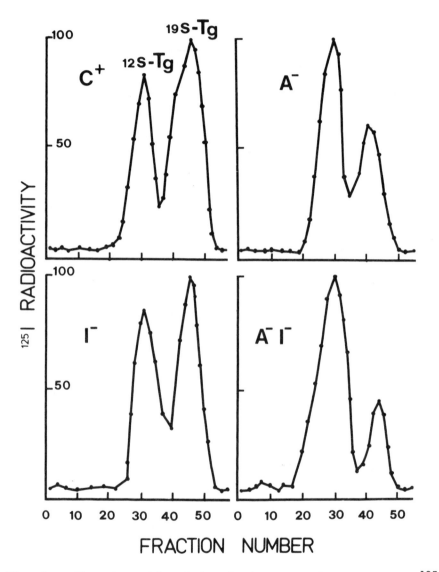

Fig. 3 : Chromatographic study of the noncovalent bonds of ^{125}I-Tg after guanidine 6M incubation. Individual values obtained for each peak are recorded by integrating the area of the normalized profile, and are shown in Table V. Pure vitamin A-deficiency (A⁻) entails more destabilizing effects than pure iodine shortage (I⁻), but the combined A⁻I⁻ deficit deter-mines the most pronounced denaturation.

Table VI : Chromatographic study of the Tg covalent bonds

Rat groups (n = 7)	Iodine atoms/ mole Tg	3-8 S compound	Monomeric ^{125}I-12S-Tg	Dimeric ^{125}I-19S-Tg
C^+	35	9	53	38
A^-	34	5	95	--
I^-	16	7	71	22
A^-I^-	13	11	89	--

last A^-I^- group manifested a moderate elevation of the MIT peak value, an even less marked rise of the DIT compound, an unchanged level of T_3, and a sharp fall of T_4. This situation generates a high MIT/DIT ratio in both vitamin A-depleted groups, whereas the T_3/T_4 ratio manifests a diverging evolution as compared to the control ratio.

DISCUSSION

Tg is an iodinated glycoprotein that serves as a substrate for thyroid hormone synthesis and storage in the follicular lumen of the endocrine organ. Tg is one of the largest glycoproteins identified in the animal tissues and characterized by striking chemical and structural similarities between several mammalian species (Spiro & Spiro, 1965; Tarutani & Ui, 1969; Arima et al., 1972; Rolland & Lissitzky, 1976; Marriq et al., 1977; Spiro, 1977a). It is generally agreed that the full-grown Tg has a molecular weight of 660,000 daltons and 19S as sedimentation coefficient. This mature Tg species constitutes about 80% of all thyroid proteins (Rall et al., 1960) and has a dimeric structure comprising two non-identical 12S-Tg subunits with molecular weight close to 330,000 daltons (van der Walt et al., 1978). The nearly similar 12S-Tg monomeric subunits are regarded as the fundamental

Table VII : Distribution of the ^{125}I-radiothyronines from purified 19S-Tg
after pronase digestion

Rat groups (n = 7)	Iodine atoms/ mole Tg	MIT	DIT	T_3	T_4	MIT/DIT ratio	T_3/T_4 ratio
C^+	35	20.6	47.5	3.5	28.4	0.43	0.12
A^-	34	32.4	60.9	1.7	21.0	0.53	0.08
I^-	16	42.9	31.3	4.7	21.1	1.37	0.22
A^-I^-	13	29.2	55.9	3.1	11.8	0.52	0.26

Each value is the mean of two measurements made on pooled fractions.

Results are expressed as percentages of the total radioactivity bound to 19S-Tg.

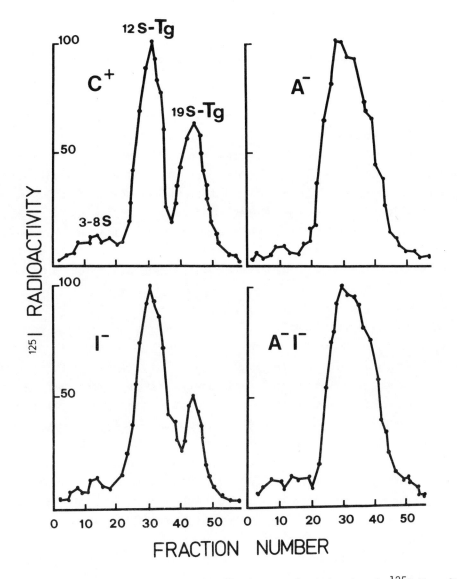

Fig. 4 : Chromatographic study of the covalent bonds of ^{125}I-Tg after guanidine treatment and dithiothreitol incubation. Individual values obtained for each peak are recorded by integrating the areas of the normalized profile, and are shown in Table VI. This denaturizing procedure transforms all the dimeric 19S-Tg into monomeric 12S-Tg species. During the subsequent centrifugation, some 12S-Tg subunits may be partially reoxidized in the C$^+$ and I$^-$ groups, leading to restoration of some 19S-Tg molecules. Such a reassembly is not observed in both A$^-$ and A$^-$I$^-$ groups, indicating that the molecular abnormalities induced by vitamin A-deficit are irreversible.

polypeptide backbone synthesized by the thyroid gland, that are held
together partly by noncovalent links (Andreoli et al., 1969; Rolland &
Lissitzky, 1972) and by covalent disulphide bonds (de Crombrugghe et al.,
1966; Pitt-Rivers & Schwartz, 1967).

The primary amino acid sequence of Tg has been partially identified
(Spiro, 1970; Lissitzky et al., 1975; Chernoff & Rawitch, 1981; Dunn et al.,
1982). Each 19S-Tg molecule possesses about 240 half-cysteine residues
allowing the closure of 120 disulphide bonds (de Crombrugghe et al., 1966;
Salvatore & Edelhoch, 1973). Almost all these S-S bridges are buried inside
the core of each 12S-Tg monomeric species (intrachain bridges), and very few
S-S linkages participate in the covalent aggregation of the two 12S monomers
(interchain bridges). Each 19S-Tg molecule contains around 140 tyrosine
residues and, under normal iodination level, about 35 to 40 iodine atoms
(Salvatore & Edelhoch, 1973).

These findings imply that only 7-10 tyrosine residues undergo coupling
reactions and iodination under the thyroid peroxydase control (Nunez &
Pommier, 1982), leading to the generation of 3 thyroxine and 0.3 triidothyronine molecules per mole of Tg and showing that only very confined and
selected areas of the Tg molecule are involved in the formation of thyroid
hormones. Up to now, it was believed that the degree of Tg iodination serves
as the main, if not the sole factor responsible for the stabilization of the
mature 19S species, since numerous experiments have shown that poorly
iodinated molecule is more prone to dissociation into the two monomeric 12S
subunits. Conversely, more highly iodinated 19S-Tg exhibits greater stability, which seems related to the closure, by halogenation, of all disulphide
bonds, though one or two remain in the free form (Nunez et al., 1966;
Andreoli et al., 1969; Edelhoch et al., 1969; Tarutani & Ui, 1969; Haeberli
et al., 1975).

The carbohydrate content of the full 19S-Tg represents about 10% of the
molecule by weight. At least six different sugars, which are incorporated in
Tg along a stepwise progression, have been identified in the carbohydrate
chains : mannose, N-acetylglucosamine, galactose, fucose, galactosamine, and
N-acetylneuraminic acid (sialic acid). Two main A and B polysaccharidic

chains have been discovered in most mammalian species, although some differences in the length and chemical composition are reported from one species
to the other (Spiro & Spiro, 1965; McQuillan & Trikojus, 1972). There is
general acceptance that these A and B chains are transferred "en bloc" from
their dolichyl pyrophosphoryl carriers onto the polypeptide Tg matrix to
which they remain attached by retinol-independent N-asparaginyl type of
linkage (De Luca, 1977; Wolf, 1977; Struck & Lennarz, 1980). Human and
guinea-pig Tg are characterized by an additional C unit rich in galactosamine (Arima et al., 1972; Spiro, 1977a). Finally, a last D unit with high
glucuronic acid content has been described in the human Tg (Spiro, 1977b).
In these two C and D fractions, the sugar chains are bound to serine or
threonine residues by an alkali-labile O-glycosidic type of linkage
suggesting that incorporation of the carbohydrates into the core of the
growing peptide necessitates the prerequisite formation of a retinol-linked
sugar acting as a direct purveyor (De Luca et al., 1973; Wolf, 1977; Rosso
et al., 1977; De Luca, 1977; Masushige et al., 1978).

Our study demonstrates that two months of vitamin A restriction lowers
plasma retinol levels, but does not affect thyroid size. Histologic alterations have been described under vitamin A-deprivation both in light and
electron microscopic sections (Coplan & Sampson, 1935; Strum, 1979). Our
results showing significantly increased TT_4, FT_4, TT_3 and FT_3 values
are in broad agreement with those reported by other workers (Garcin &
Higueret, 1977; Morley et al., 1978; Oba & Kimura, 1980; Garcin & Higueret,
1980). This situation indicates a biochemical hyperthyroidism which is not
accompanied by the expected TSH downregulation that would occur if the
negative feedback control of the pituitary secretion was preserved. In fact,
vitamin A-depleted rats manifest an appropriate hypophyseal response to
exogenous TRH administration (Morley et al., 1978; Garcin & Higueret, 1979),
implying that TSH synthesis, storage and release is normal. Since the rats
exhibit neither the clinical nor biological symptoms of hyperthyroidism, the
paradoxical association of high levels of circulating total and free thyroid
hormones and normal TSH values suggests a generalized unresponsiveness of
the peripheral organs, including the pituitary, to the action of the thyroid
hormones. This partial insensitivity appears as a salient hallmark of
vitamin A-deprivation and is best explained by a defect situated at the

target site of hormonal action (Refetoff, 1982). We anticipate that an impaired retinol-dependent glycosylation of the intracellular receptors, hindering their ability to interact with T_3, could be the molecular anomaly responsible for tissue refractoriness.

The elevated TT_4, FT_4, TT_3 and FT_3 concentrations in vitamin A-shortage can be ascribed to a combination of at least three causative factors : First, higher Tg degradation rate, with the increase in release of hormonal compounds by the thyroid gland. A comparable situation is documented under several endocrine disorders, among which are thyroiditis and goitre, where the high circulating levels of iodoproteins may be regarded as a nonspecific consequence of glandular injury (Pittman & Pittman, 1966; Salvatore & Edelhoch, 1973; Glinoer et al., 1974). More rapid Tg turnover in vitamin A-deficient rats is strongly suggested by microscopic studies showing accelerated entrance of follicular cells into the colloid lumen, together with an unusual abundance of lysosomes (Strum, 1979). Second, abnormalities in the peripheral T_4 to T_3 monodeiodinating pathway. The elevated rT_3 blood values after vitamin A restriction support this idea. Blockade of the normal dehalogenation reaction can also induce high blood TT_4 values (Jansen et al., 1982). The hepatic conversion occurs at the microsomal level and requires an adequate supply of at least two cofactors, reduced glutathione (GSH) and reduced nicotinamide adenine dinucleotide phosphate (NADPH) (Balsam & Ingbar, 1979). The latter compound is depressed during vitamin A-depletion (Blaizot & Bonmort, 1979), whereas a decreased level of the former is to be suspected, since its addition to liver microsomal preparations improves the lost deiodinating activity (Higueret & Garcin, 1982). Third, upstream retention of thyroid hormones in extracellular spaces due to peripheral resistance. As a matter of fact, T_4 circulates largely in the extracellular fluids, in contrast to T_3 which is mainly confined to the intracellular compartment (Inada et al., 1975). The observation in vitamin A-depleted rats of an increased T_4 biological half-life, associated with decreased T_3 distribution space (Higueret & Garcin, 1982) sustains this view. Apparently, the specific carrier proteins are not involved in the observed abnormalities, since the concentration of serum prealbumin, the major transport protein for the thyroid hormones in the rat, is not depressed under vitamin A-deficiency (Navab et al., 1977).

However, subtle alterations in the binding affinities of prealbumin may occur. This is suggested by observations that, on electrophoresis, labelled T_4 and T_3 shift towards a new postalbumin zone (Garcin & Higueret, 1980; Oba & Kimura, 1980). The physiological significance of this augmented TBG*-like area under vitamin A restriction awaits elucidation.

The blood parameters recorded in the I^- group are similar to those reported in previous clinical and experimental works (Studer & Greer, 1968; Delange, 1974; Ingenbleek et al., 1980). The thyroid-pituitary negative feedback is here fully effective, and the low FT_4 circulating levels appear to be the best peripheral indicator of the thyroid status and the sensor modulating TSH secretion (Ingenbleek, 1980). High TSH production exerts direct trophic effects on the thyroid tissue (Dumont et al., 1978), enhances its avidity for the rarefied halogen (Inoue & Taurog, 1968), and specifically stimulates the activity of the thyroid 5'-monodeiodinase (Erickson et al., 1982). The A^-I^- group is characterized by blood parameters situated at an intermediate level between the two extreme A^- and I^- poles.

The incubation of the rat thyroids in the presence of tritiated mannose demonstrates that the RPM mannolipid fraction really exists in measurable amounts in the rat thyroid cytosol (Table III). Moreover, our data show a 50 to 60% reduction in endogenous RPM synthesis under both A^- and A^-I^- dietary shortage. On the contrary, values obtained for DPM are slightly elevated in the vitamin A-depleted rat thyroid tissue. A comparable accumulation of the dolichol systems has been reported in vitamin A-deprived rat liver (Rosso et al., 1981). Taken together, these findings suggest that the decrease in RPM synthesis at cytosolic level is the direct result of vitamin A-deficit, and that the higher DPM concentrations could be the consequence of less efficient utilization of the latter mannolipid fractions for further protein glycosylations. It was recently postulated that both RPM and DPM mannolipid intermediates could be interconnected by a converging

---

* TBG : Thyrobinding Globulin

pathway which could be affected in opposite directions under vitamin A-deprivation (De Luca et al., 1982). This view is consistent with a regulatory role played by RPM on the subsequent N-glycosylation reactions.

The reduced incorporation of ^{3}H-mannose into the core of both A^- and A^-I^- thyroglobulin species (Table IV) indicates that the primary sequence of the growing peptides is modified under vitamin A-deficient conditions. Of the utmost importance is the fact that the depressed Tg mannosylation is exactly of the same order of magnitude as the decreased RPM synthesis. This similarly low recovery rate obtained for ^{3}H-mannose by two independent technical approaches supports the concept that the uptake of this sugar by native Tg is governed by the law of mass action and, through the mediation of RPM supply, directly influenced by vitamin A availability. Moreover, and although this point lacks final proof to date, this survey suggests that the rat Tg matrix may contain at least one retinol-sensitive O-mannosylated chain comparable to the C and D units described in man and guinea-pig (Arima et al., 1972; Spiro, 1977a; Spiro, 1977b). Obviously, pure iodine shortage (I^- group) by no means interferes at any level neither with the endogenous synthesis of RPM or DPM, nor with the insertion of ^{3}H-mannose into the Tg polypeptide chain.

The study of the 19S-Tg noncovalent and covalent bonds (Tables V & VI and Fig. 3 & 4) clearly reveals that vitamin A-deficit may entail per se, and regardless of iodination level, destabilizing effects similar to those triggered by pure iodine deprivation. These structural changes may be assigned to the persistent opening of several normally closed disulphide bonds. This crucial point may be ascertained by the fact that the Tg found in both C^+ and A^- groups is characterized by the same normal halogenation degree (34-35 atoms I^-/mole) indicating that iodine is not implicated in the observed anomalies. The combination of both A^- and I^- deficits produced the most pronounced disturbances in Tg conformation, showing that iodine restriction superimposed on vitamin A-deficit aggravates the glyco-peptide dysmaturation. These experiments also display that pure iodine deficiency allows partial reoxidizing of some cystein residues, leading to reassembly of some reduced 12S monomers into 19S-Tg molecules. This reaggregation did not occur in the A^- and A^-I^- groups, implying that the molecular alterations induced by vitamin A-deficiency have become irreversible.

The profile of the iodinated fractions found on the Tg matrix depends on the nutritional environment (Table VII). In the A^- group, production of both T_3 and T_4 hormonal compounds is inhibited by 50% and by 25%, respectively, in contrast to higher concentrations of MIT and DIT residues. This unusual pattern is, here too, exaggerated by combined A^-I^- deficit. An entirely different Tg halogenation spectrum is recorded in the I^- group, where MIT and T_3 synthesis is clearly enhanced at the expense of depressed DIT and T_4 generation. These findings explain why both MIT/DIT and T_3/T_4 ratios are increased under iodine shortage, whereas the former ratio is elevated and the latter diminished as a result of vitamin A deprivation. Summing up, it is apparent that the goitrous hypertrophy induced by iodine deprivation may be regarded as an adaptative dysfunction in which the thyroid-pituitary negative feedback and the coupling reactions between iodotyrosine residues intervene with normal or even with improved efficiency. High TT_3 and FT_3 together with low TT_4 and FT_4 blood levels may be related to the specific activation of the thyroid 5'-mono-deiodinating enzyme which is directly controlled by TSH. Indeed, all regulatory mechanisms characterizing normal thyroid function are reset at new (higher or lower) thresholds of activity, in order to design an iodide-sparing scenario that is beneficial to the general thyroid economy. This iodide-dependent goitrogenic process is a late event in thyroglobulin maturation, since iodination is shown to occur at the apical edge of the follicular cell (Ekholm & Wollman, 1975) when the synthesis of the Tg glyco-peptide matrix is fully achieved (Edelhoch et al., 1973; Haeberli et al., 1975).

In contrast, vitamin A-deficiency very early alters the thyroglobulin structure. Studies using labelled saccharides have shown that mannose is already incorporated at ribosomal level, initiating most probably the stepwise Tg glycosylation at the 3-8 S peptide level (Whur et al., 1969; Vecchio et al., 1971; Bouchilloux et al., 1973). Furthermore, mannose is recognized as capable of determining significant changes in the steric configuration of other glycoproteins, such as immunoglobulins (Sutherland et al., 1972; Kaverzneva & Shmakova, 1981). We assume, therefore, that the abnormal primary sequence of poorly mannosylated Tg modifies the glyco-peptide tertiary structure in such a way that its hormonogenic potency is severely depressed. Under physiological conditions, only a restricted number of tyrosine residues undergoes iodination and coupling reaction with optimal efficacy, a situation indicating that the amino acids involved in the vicinity of the forming sites are submitted to stringent sequential constraints (Chernoff & Rawitch, 1981; Dunn et al., 1982; Nunez & Pommier, 1982). Our data collected for Tg noncovalent and covalent bonds, together with its iodination profile, strongly suggest that the undermannosylated iodoprotein is characterized by an unpropritious spatial rearrangement of these amino acids. This incorrect alignment apparently increases the distance separating many residues, hampers the normal closure of several disulphide linkages, rendering the tyrosine-tyrosine coupling reactions less efficient. The vitamin A-dependent goitrogenic process may thus be regarded as the result of a molecular defect. It becomes obvious that the permanent antagonism between retinol and iodine on thyroid function is mainly focused at the level of T_3 synthesis, since the MIT-DIT coupling reaction appears to be markedly inhibited under the vitamin-deficit, and highly enhanced in the halogen shortage.

The question now arises whether these experimental findings have any implications in human goitrous disease. Certainly, total iodine restriction and total vitamin A deprivation represent two extreme poles which are not compatible with prolonged survival of human beings. A growing body of data, however, indicates that populations living under poor sanitary conditions, especially in developing countries, may be simultaneously affected by diverse nutritional disorders of mild or moderate severity. Moreover, the varying pathogenic aspects of these deprivation illnesses may be closely interwoven. Fifteen years ago, György (1968) demonstrated that protein- and vitamin A-depleted patients had strikingly similar abnormalities of their electroretinograms, indicating that both deficits may lead to comparable ocular side consequences. This clinical observation was later substantiated by biological investigations showing that normal subjects, malnourished and goitrous patients reveal, regardless of age, sex and social environment, blood concentrations for the three components of the retinol circulating complex which remain attached in an equimolar 1:1:1 ratio (Ingenbleek, 1980; Ingenbleek et al., 1980). The persistence of this high positive correlation implies that the synthesis rate of prealbumin by the liver determines the peripheral retinol status of the body (Ingenbleek et al., 1975a; 1975b). Since the decline of plasma prealbumin is usually a very sensitive indicator of protein malnutrition (Ingenbleek et al., 1972; Ingenbleek et al., 1975a), it is understandable that chronic protein-deficit replicates at cellular level the adverse effects generated by insufficient intake of vitamin A. Results obtained with our animal model thus appear extrapolable to human malnutrition and provide, in combination with iodine deficiency, valid explanations for the unclarified epidemiological findings recorded in endemic goitrous areas (Ingenbleek & De Visscher, 1979).

REFERENCES

Adler, H.L. & Roessler E.B. (1968) Probability and Statistics. Freeman Press, San Francisco, USA.

Andreoli, M., Sena, L., Edelhoch, H. & Salvatore, G. (1969) The noncovalent subunit structure of human thyroglobulin. Arch. Biochem. Biophys. 134: 242-248.

Anonymous (1924) Relation of goitre to the iodine ingested. Brit. Med. J. 2:528-529.

Arima, T., Spiro, M.J. & Spiro, R.G. (1972) Studies on the carbohydrate units of thyroglobulin. Evaluation of their microheterogeneity in the human and calf proteins. J. Biol. Chem. 247:1825-1835.

Balsam, A. & Ingbar, S.H. (1979) Observations of the factors that control the generation of triiodothyronine from thyroxine in rat liver and the nature of the defect induced by fasting. J. Clin. Invest. 63:1145-1156.

Baumann, E. (1895) Ueber das normale Vorkommen von Jod im Tierkorpfe. Hoppe-Seyler's. Z. Physiol. Chem. 21:319-330.

Benotti, J., Benotti, N., Pino, S. & Gardyna, H. (1965) Determination of total iodine in urine, stool, diets and tissue. Clin. Chem. 11:392-396.

Blaizot, S. & Bonmort, J. (1979) Alimentation en temps limité ("meal-feeding") et avitaminose A : Echanges respiratoires et adaptation métabolique du rat. Ann. Nutr. Aliment. 33:363-383.

Bouchilloux, S., Chabaud, O. & Ronin, C. (1973) Cell-free peptide synthesis and carbohydrate incorporation by various thyroid particles. Biochim. Biophys. Acta 322:401-420.

Chatin, A. (1853) Un fait dans la question du goitre et du crétinisme. C.R. Acad. Sci. (Paris) 36:652-654.

Chernoff, S.B. & Rawitch, A.B. (1981) Thyroglobulin structure-function. Isolation and characterization of a thyroxine-containing polypeptide from bovine thyroglobulin. J. Biol. Chem. 256:9425-9430.

Coplan, H.M. & Sampson, M.M. (1935) The effects of a deficiency of iodine and vitamin A on the thyroid gland of the albino rat. J. Nutr. 9:469-487.

de Crombrugghe, B., Pitt-Rivers, R. & Edelhoch, H. (1966) The properties of thyroglobulin. XI. The reduction of disulfide bonds. J. Biol. Chem. 241: 2766-2773.

Delange, F. (1974) Endemic Goitre and Thyroid Function in Central Africa. Monographs in Paediatrics (Ed. Karger S., Basel), vol. 2.

De Luca, L.M., Maestri, N., Rosso, G.C. & Wolf, G. (1973) Retinol glycolipids. J. Biol. Chem. 248:641-648.

De Luca, L.M. (1977) The direct involvement of vitamin A in glycosyl transfer reactions of mammalian membranes. Vitam. Horm. 35:1-57.

De Luca, L.M., Brugh, M.R., Silverman-Jones, C.S. & Shidoji, Y. (1982) Synthesis of retinyl phosphate mannose and dolichyl phosphate mannose from endogenous and exogenous retinyl phosphate and dolichyl phosphate in microsomal fraction. Specific decrease in endogenous retinyl phosphate mannose synthesis in vitamin A deficiency . Biochem. J. 268:159-170.

Dumont, J.E., Boeynaems, J.M., Decoster, C., Erneux, C., Lamy, F., Lecocq, R., Mockel, J., Unger, J. & Van Sande, J. (1978). Biochemical mechanisms in the control of thyroid function and growth. Adv. Cyclic Nucleotide Res. 9:723-734.

Dunn, J.T., Kim, P.S. & Dunn, A.D. (1982) Favored sites for thyroid hormone formation on the peptide chains of human thyroglobulin. J. Biol. Chem. 257:88-94.

Edelhoch, H., Carlomagno, M.S. & Salvatore, G. (1969) Iodine and the structure of thyroglobulin. Arch. Biochem. Biophys. 134: 264-265.

Ekholm, R. & Wollman, S.H. (1975) Site of iodination in the rat thyroid gland deduced from electron microscopic autoradiographs. Endocrinology 97:1432-1444.

Erickson, V.J., Cavalieri, R.R. & Rosenberg, L.L. (1982) Thyroxine-5'-deiodinase of rat thyroid, but not that of liver, is dependent on thyrotropin. Endocrinology 111:434-440.

Finucane, J.F. & Griffiths, R.S. (1976) A rapid and simple method for simultaneous measurement of serum free thyroxine and triiodothyronine fractions. J. Clin. Pathol. 29:949-954.

Gaitan, E. (1980) Goitrogens in the etiology of endemic goiter. In : Endemic Goiter and Endemic Cretinism (Ed. Stanbury, J.B. & Hetzel, B.S.) John Wiley & Sons Inc., New York, pp. 219-236.

Garcin, H. & Higueret, P. (1977) Influence de la carence en vitamine A sur la thyroxinémie du rat. C.R. Acad. Sci. (Paris) 285:531-533.

Garcin, H. & Higueret, P. (1979) Vitamin A-deficiency and thyrotropin secretion in the rat. J. Physiol. (Paris) 75:887-890.

Garcin, H. & Higueret, P. (1980) Free and protein-bound triiodothyronine in the serum of vitamin A-deficient rats. J. Endocrinol. 84:135-140.

Garry, P.J., Pollack, J.D. & Owen, G.M. (1970) Plasma vitamin A assay by fluorometry and use of a silicic acid column technique. Clin. Chem. 16:766-772.

Glinoer, D., Puttemans, N., Van Herle, A.J., Camus, M. & Ermans, A.M. (1974) Sequential study of the impairment of thyroid function in the early stage of subacute thyroiditis. Acta Endocrinol. (Kbn) 77:26-34.

György, P. (1968) Protein-calorie and vitamin A malnutrition in Southeast Asia. Fed. Proc. 27:949-953.

Haeberli, A., Salvatore, G., Edelhoch, H. & Rall, J.E. (1975) Relationship between iodination and polypeptide chain composition of thyroglobulin. J. Biol. Chem. 250:7836-7841.

Higueret, P. & Garcin, H. (1982) Peripheral metabolism of thyroid hormones in vitamin A-deficient rats. Ann. Nutr. Metab. 26:191-200.

Inada, M., Kasagi, K., Kurata, S., Kazama, Y., Takayama, H., Torizuka, K., Fukase, M. & Soma, T. (1975) Estimation of thyroxine and triidothyronine distribution and of the conversion rate of thyroxine and triiodothyronine in man. J. Clin. Invest. 55:1337-1348.

Ingenbleek, Y., De Visscher, M. & De Nayer, Ph. (1972) Measurement of pre-albumin as index of protein-calorie malnutrition. Lancet 2:106-109.

Ingenbleek, Y., Van den Schrieck, H.G., De Nayer, Ph. & De Visscher, M. (1975a) Albumin, transferrin and the thyroxine-binding prealbumin/retinol-binding protein (TBPA-RBP) complex in assessment of malnutrition. Clin. Chim. Acta 63:61-67.

Ingenbleek, Y., Van den Schrieck, H.G., De Nayer, Ph., and De Visscher, M. (1975b) The role of retinol-binding protein in protein-calorie malnutrition. Metabolism 24:633-641.

Ingenbleek, Y., & De Visscher, M. (1979) Hormonal and status : Critical conditions for endemic goiter epidemiology ? Metabolism, 28:9-19.

Ingenbleek, Y., Luypaert, B. & De Nayer, Ph. (1980) Nutritional status and endemic goitre. Lancet 1:388-392.

Ingenbleek, Y. (1980) Thyroid function in nonthyroid illnesses. In : The Thyroid Gland (Ed. De Visscher, M.) Raven Press, New York, pp. 499-527.

Inoue, K. & Taurog, A. (1968) Acute and chronic effects of iodide on thyroid radioiodine metabolism in iodine-deficient rats. Endocrinology 83:279-290.

Jansen, M., Krenning, E.P., Oostdijk, W., Docter, R., Kingma, B.E., van den Brande, J.V. & Henneman, G. (1982) Hyperthyroxinaemia due to decreased peripheral triiodothyronine production. Lancet 2:849-852.

Kaverzneva, E.D. & Shmakova, F.V. (1981) Role of carbohydrates in immunoglobulin M. VII. Implication of mannose-rich oligosaccharide chains in acquisition of immunoglobulin spatial structure. Bioorg. Khim. 7:1254-1260.

Kelly, F.C. & Snedden, W.W. (1960) Prevalence and geographical distribution of endemic goitre. In : Endemic Goitre, Monograph Series No 44, WHO, Geneva, pp. 27-233.

Kiorpes, T.C., Molica, S.J. & Wolf, G. (1976) A plasma glycoprotein depressed in vitamin A deficiency in the rat : Alpha-1-macroglobulin. J. Nutr. 106:1659-1667.

Koutras, D.A., Christakis, G., Trichopoulos, D., Dakou Voutetaki, A., Kyriakopoulos, V., Fontanares, P., Livadas, D.P., Gatsios, D. & Malamos, B. (1973) Endemic goiter in Greece : Nutritional status, growth, and skeletal development of goitrous and nongoitrous populations. Am. J. Clin. Nutr. 23:1360-1368.

Koutras, D.A. (1980) Trace elements, genetic and other factors. In : Endemic Goiter and Endemic Cretinism (Ed. Stanbury, J.B. & Hetzel, B.S.) John Wiley & Sons Inc., New York, pp. 255-268.

Lissitzky, S., Mauchamp, J., Reynaud, J. & Rolland, M. (1975). The constituent polypeptide chain of porcine thyroglobulin. FEBS Lett. 60:359-363.

Lowry, O.H., Rosebrough, N.J., Farr, A.L., & Randall, R.J. (1951) Protein measurement with the Folin phenol agent. J. Biol. Chem. 193:265-275.

Maffoni, A.C. (1846) Brevi cenni sullo stato sanitario del Piemonte con proposta di alcuni mezzi. Atti Reale Accad. Medicochir. Torino 2:453-484.

Marine, D. & Kimball, O.P. (1920) Prevention of single goiter in man. Fourth paper. Arch. Int. Med. 25:661-672.

Marriq, C., Rolland, M. & Lissitzky, S. (1977) Polypeptide chains of 19S thyroglobulin from several mammalian species and of porcine 27S iodoprotein. Eur. J. Biochem. 79:143-148.

Masushige, S., Schreiber, J.B. & Wolf, G. (1978) Identification and characterization of mannosyl retinyl phosphate occurring in rat liver and intestine in vivo. J. Lipid Res. 19:619-627.

McCullagh, S.F. (1963) The Huon peninsula endemic. I. The effectiveness of an intramuscular depot of iodized oil in the control of endemic goitre. Med. J. Austral. 1:769-777.

McQuillan, M.T. & Trikojus, V.M. (1972) Thyroglobulin. In : Glycoproteins, their Composition, Structure and Function (Ed. Gottschalk, A.) Elsevier, Amsterdam, pp. 926-963.

Medeiros-Neto, G.A. (1980) General nutrition and endemic goiter. In : Endemic Goiter and Endemic Cretinism (Ed. Stanbury, J.B. & Hetzel, B.S.) John Wiley & Sons Inc., New York, pp. 269-283.

Morley, J.E., Damassa, D.A., Gordon, J., Pekary, A.E. & Hershman, J.M. (1978) Thyroid function and vitamin A deficiency. Life Sci. 22:1901-1906.

Navab, M., Smith, J.E., Goodman, D.S. (1977) Rat plasma prealbumin : Metabolic studies on effects of vitamin A status and on tissue distribution. J. Biol. Chem. 252:5107-5114.

Nunez, J., Mauchamp, J., Pommier, J., Circovik, T. & Roche, J. (1966) Relationship between iodination and conformation of thyroglobulin. Biochem. Biophys. Res. Comm. 23:761-768.

Nunez, J. & Pommier, J. (1982) Formation of thyroid hormones. Vitam. Horm. 39:175-229.

Oba, K. & Kimura, S. (1980) Effects of vitamin A deficiency on thyroid function and serum thyroxine levels in the rat. J. Nutr. Sci. Vitaminol. 26:327-334.

Osborn, R.H. & Simpson, T.H. (1968) Gel filtration of iodoamino acids. J. Chromatogr. 34:110-111.

Pittman, C.S. & Pittman, J.A. Jr (1966) A study of the thyroglobulin, thyroidal protease and iodoproteins in two congenital goitrous cretins. Am. J. Med. 40:49-57.

Pitt-Rivers, R. & Schwartz, H.L. (1967) The thiol groups of thyroglobulin. Biochem. J. 105:28-29.

Rall, J.E., Robbins, J. & Edelhoch, H. (1960) Iodoproteins in the thyroid. Ann. N.Y. Acad. Sci. 86:373-399.

Refetoff, S. (1982) Syndromes of thyroid hormone resistance. Am. J. Physiol. 243:E88-E98.

Rolland, M. & Lissitzky, S. (1972) Polypeptides non-covalently associated in 19S thyroglobulin. Biochim. Biophys. Acta 278:316-336.

Rolland, M. & Lissitzky, S. (1976) Endogenous proteolytic activity and constituent polypeptide chains of sheep and pig 19S thyroglobulin. Biochim. Biophys. Acta 427:696-707.

Rosso, G.C., Masushige, S., Quill, H. & Wolf, G. (1977) Transfer of mannose from mannosylretinylphosphate to protein. Proc. Natl. Acad. Sci. USA 74, 3762-3766.

Rosso, G.C., Bendrick, C.J. & Wolf, G. (1981). In vivo synthesis of lipid-linked oligosaccharides in the livers of normal and vitamin A-deficient rats. J. Biol. Chem. 256:8341-8347.

Salvatore, G. & Edelhoch, H. (1973) Chemistry and biosynthesis of thyroid iodoproteins. In : Hormonal Proteins and Peptides. Vol. 1. (Ed. Li, C.H.) Academic Press, New York, pp. 201-241.

Silverman-Jones, C.S., Frot-Coutaz, J.P. & De Luca, L.M. (1976) Separation of mannosylretinylphosphate from dolichylmannosylphosphate by solvent extraction. Anal. Biochem. 75:664-667.

Sorimachi, K. & Ui, N. (1974) Comparison of the iodoaminoacid distribution in thyroglobulins obtained from various animal species. Gen. Comp. Endocrinol. 24:38-43.

Spiro, R.G. & Spiro, M.J. (1965) The carbohydrate composition of the thyroglobulins from several species. J. Biol. Chem. 240:997-1001.

Spiro, M.J. (1970) Studies on the protein portion of thyroglobulin. Amino acid compositions and terminal amino acids of several thyroglobulins. J. Biol. Chem. 245:5820-5826.

Spiro, M.J. (1977a) Guinea pig thyroglobulin : Molecular weight of subunits and amino acid and sugar composition. Biochem. Biophys. Res. Comm. 77: 874-882.

Spiro, M.J. (1977b) Presence of a glucuronic acid-containing carbohydrate unit in human thyroglobulin. J. Biol. Chem. 252:5424-5430.

Stanbury, J.B. (1969) Preface. In : Endemic Goiter. PAHO Publication No 193 (Ed. Stanbury, J.B.), Washington, p. vii.

Stanbury, J.B. (1973) Factors which may alter the epidemiology of endemic goiter. Acta Endocrinol. (Kbn) (Suppl. 179) 74:9-10.

Struck, D.K. & Lennarz, W.J. (1980) The function of saccharide-lipids in synthesis of glycoproteins. In : Biochemistry of Glycoproteins and Proteoglycans (Ed. Lennarz, W.J.) Plenum Press, New York, pp. 35-83.

Strum, J.M. (1979) Alterations within the thyroid gland during vitamin A deficiency. Am. J. Anat. 156:169-181.

Studer, H. & Greer, M.A. (1968) The Regulation of Thyroid Function in Iodine Deficiency (Ed. Hans Huber, Bern) pp. 1-119.

Sutherland, E.W. 3rd, Zimmerman, D.H. & Kern, M. (1972) Synthesis and secretion of gammaglobulin by lymph node cells : The acquisition of carbohydrate residues of immunoglobulin in relation to interchain disulfide bond formation. Proc. Natl. Acad. Sci. USA 69:167-171.

Tarutani, O. & Ui, N. (1969) Subunit structure of hog thyroglobulin. Dissociation of non-iodinated and highly iodinated preparations. Biochim. Biophys. Acta 181:136-145.

Tkacz, J.S., Herscovics, A., Warren, C.D. & Jeanloz, R.W. (1974) Mannosyl transferase activity in calf pancreas microsomes. Formation from guanosine diphosphate-D-[^{14}C] mannose of a ^{14}C-labeled mannolipid with properties of dolichyl mannopyranosyl phosphate. J. Biol. Chem. 249: 6372-6381.

van der Walt, B., Kotzé, P.P., van Jaarsveld, P. & Edelhoch, H. (1978) Evidence that thyroglobulin contains nonidentical half-molecule subunits. J. Biol. Chem. 253:1853-1858.

Vecchio, G., Carlomagno, M.S. & Consiglio, E. (1971) Identification and characterization of two labeled intermediates in the biosynthesis of rat thyroglobulin. J. Biol. Chem. 246:6676-6682.

Whur, P., Herscovics, A. & Leblond, C.P. (1969) Radioautographic visualization of the incorporation of galactose-^{3}H and mannose-^{3}H by thyroid in vitro in relation to the stages of thyroglobulin synthesis. J. Cell. Biol. 43:289-311.

Wolf, G. (1977) Retinol-linked sugars in glycoprotein synthesis. Nutr. Rev. 35:97-99.

IS THE ADULT PROTEIN-ENERGY MALNUTRITION SYNDROME THE SAME AS THAT DESCRIBED IN THE INFANT ?

Jean MAURON and Ilse ANTENER

Nestlé Products Technical Assistance Co. Ltd, Research Department, CH-1814 La Tour-de-Peilz, Switzerland

Hospital of Yasa Bonga (Zaïre)

SUMMARY

Protein-energy malnutrition, a multi-factorial disease, has been described predominantly in the infant. It was the aim of this research to give a biochemical assessment of the adult form and to compare it to the infantile syndrome within the same socio-cultural context of central Zaïre (Kwilu region).

Thirty-four children, 22 women and 2 men suffering from marasmic kwashiorkor at the hospital of Yasa-Bonga (Kwilu) were submitted to a complete set of 7 anthropometric and 60 biochemical tests. The control values were taken from healthy well-fed children and adults from Yasa-Bonga; for certain parameters, rural adult control values were also obtained.

Dyspigmentation was found in all patients, children and adults alike. The other symptoms were, in decreasing order of importance : oedema, dermatitis, apathy and liver enlargement, often accompanied by associated secondary pathology. In the children, all anthropometric indices were well below normal.

In serum, total protein, albumin, prealbumin, ceruloplasmin and haemoglobin were reduced; the α_1, α_2 and β-globulins were slightly reduced in infants but not much modified in adults, whereas γ-globulins were slightly increased in adults only. IgG and IgM were increased in both infants and adults, the enhancement was less pronounced for IgA.

Essential amino acids in serum were reduced in the patients and most non-essential amino acids raised, with the exception of tyrosine and arginine which were reduced like the essential amino acids. Some ratios (phenylalanine/tyrosine, serine/threonine, and non-essential/essential amino acids) proved to be very sensitive parameters for this type of protein-energy malnutrition. No differences were found in the amino acid levels between adult and infant patients, with the exception of alanine which was higher in the adults. Alanine levels were also high in the rural adult controls as compared to the European controls, probably due to the extremely high carbohydrate (manioc) diet.

Serum electrolytes were normal in adult patients, with the exception of low Ca, whereas in children P and Mg were also low.

Total lipids and cholesterol were reduced in adult and infant patients.

The urinary excretion of all parameters measured (N, urea, creatinine, hydroxyproline, electrolytes, trace elements and some vitamins) was reduced in the patients with only small differences between adults and infants.

Starch was present in the faeces in two-thirds of the patients, disaccharides in 15% and an increased amount of glucose in 85%, showing the malabsorption syndrome.

Two of eight serum enzymes determined showed a typical pathological behaviour : γ-glutamyl-transpeptidase (γ-GT) was

enhanced and cholinesterase (CHE) was diminished in the malnourished infants and adults. The abnormal γ-GT isoenzyme pattern
was similar in infant and adult patients; especially, the
appearance of a band near albumin (A-α_1-γ-GT) carried a
very poor prognosis in both children and adults.

Another feature of our investigation was the design of dietary treatment using mainly local foods. Upon dietary rehabilitation, the parameter in serum showing the most sensitive
response was transferrin, followed by albumin and ceruloplasmin,
whereas prealbumin was very slow to react, especially in adults,
showing that the improvement of the general nutritional status
of the individual is rather sluggish. In general, the response
of the adults to the diet was slower than that of the children
but it was not qualitatively different.

In urine, increased secretion of urea and hydroxyproline were
striking, especially in children. In faeces, starch and sugars
disappeared.

The evolution of the illness during treatment is best
followed by measuring γ-GT and CHE. Concomitant decrease of
γ-GT and increase in CHE to normal levels signifies good
response to the treatment and favourable evolution. Where the
normalisation of the 2 enzyme levels cannot be obtained, the
prognosis is bad indeed. The normalisation of γ-GT isoenzymes
is also an excellent prognostic tool.

In assessing nutritional rehabilitation we tried, finally, to
distinguish between parameters responding more to the adequacy
of the diet and those more related to the state of the individual and having therefore a certain prognostic value.

Since most clinical signs and all biochemical parameters
measured were practically identical in infants and adults
suffering from malnutrition and since dietary rehabilitation,

although slower in adults, was accompanied by the same biochemical modifications, we conclude that the syndrome, described here in some detail for adults, is the same as that found in infants and does not depend in this rural environment upon the age of the individual but on a particular stress situation due to poor diet, poverty and infection, affecting the most vulnerable groups of individuals.

Lastly, the syndrome is not truly protein-energy malnutrition, but generalized malnutrition including minerals and vitamins, with a particular involvement of the liver.

* * *

It has become more and more apparent during the last decade or so that the syndrome called protein-energy malnutrition (PEM) can be compared to a chameleon, changing its aspect and expression according to the circumstances. Indeed, not only do the manifestations of the disease vary according to the relative proportions of the energy and the protein deficit, but also in relation to concomitant deficiencies in vitamins or minerals, as well as general socio-economic conditions, breast-feeding habits, and so on. In addition, the clinical and biochemical features of the syndrome depend on the hormonal situation of the individuals concerned at the onset of deprivation.

The aim of this research was to characterize PEM, as it occurs in a given socio-cultural context, namely that of the rural population in Central Zaïre (Kwilu region). A typical feature of the malnutrition in that region is that it affects not only the preschool child, but also, frequently, adult lactating women. The local name for PEM is "mbwaki" meaning "copper-colour". We tried to make an integrated approach to the malnutrition situation, combining nutrition surveys among the healthy rural population in the nearby villages, with the biochemical assessment of malnutrition in the Hospital of Yasa-Bonga in order to design dietary habilitation schemes for the malnourished patients.

Our presentation will deal essentially with the biochemical assessment of malnutrition and the comparison of infant and adult PEM.

We shall start with a very superficial view of the general nutrition situation in the region showing the result of a nutrition survey made on a family basis by the direct weighing method and analysis of aliquots of all the food eaten (Tables I and II). The tentative results shown are from the Kumbimashi family in the village of Mikula. Because of extremely uniform food habits, these results are quite representative for the other families in the region and are shown here only as "pars pro toto". The complete results, including other families, will be published elsewhere.

The main staple of these subsistence farmers is manioc, covering almost 90% of the total calories. The calorie requirements are just covered, whereas protein supply is insufficient, as is the supply of vitamins B_2, A, B_1 and Cu. The intake of the other nutrients is adequate or at least borderline, like that of zinc. The population adapts to the low protein supply by reducing body weight and size. In any stressing situation, the vulnerable groups, namely the pre-school children and the lactating women, are prompt to develop PEM. In children, the stress is generally the displacement from the breast, and infections such as measles; in the adult women, the main cause for malnutrition is socio-economic deprivation (unmarried mothers, abandoned wives), combined with frequent pregnancies and prolonged lactation. A rather typical situation is to find a mother with early PEM symptoms, suckling a still normal child, accompanied by an older child that has been displaced from the breast and is, therefore, suffering from PEM like its mother.

Marked seasonal variations are observed, malnutrition being most frequent in the fall, after the short dry season, before the new crop can be harvested in January.

There is little doubt that the relatively high frequency of adult PEM in that region is linked to the extremely high manioc consumption and the need to rely totally on the food produced on the poor, acid soils or gathered in the forest.

With this general background in mind, we shall now describe the two forms of PEM observed in that region.

ASSESSMENT OF MALNUTRITION[a]

Patients and clinical symptoms

Between 1971 and 1975 (at four different periods), we studied 56 patients comprising 33 children (aged 1 to 9), 1 adolescent, 20 women (aged 19 to 25) and 2 men (aged between 40 and 50).

69.7% of the children involved were aged between 3 and 6. This represents wider age limits than those usually described for kwashiorkor (Pereira et al., 1974) but similar to those of the patients observed by DeMaeyer et al. (1958) in Kivu. Most of the women studied were multiparous lactating women.

The clinical symptoms and their frequency are represented in Fig. 1. Dyspigmentation is found in all patients. It affects the hair, eyebrows and eyes-lashes, and then, in the most serious cases, spreads across the whole body, starting with the extremities. The other symptoms were in decreasing order of importance : oedema, dermatosis, apathy and to a lesser extent, liver enlargement. Apart from these typical signs of malnutrition, we find the associated pathology often linked in some way or other to malnutrition in tropical countries (diarrhoea, virus infections, anaemia).

Anthropometric measurements (Fig. 2-4)

The classification of the children according to weight/age, height/age and thorax/age shows that all are well below normal, if we use the Boston standard (Nelson, 1964). It should be noted that the weight/age and thorax/age relations are even more abnormal than the height/age relation.

(a) For a detailed description of the methods and discussion of the results see : Antener et al., 1977a.

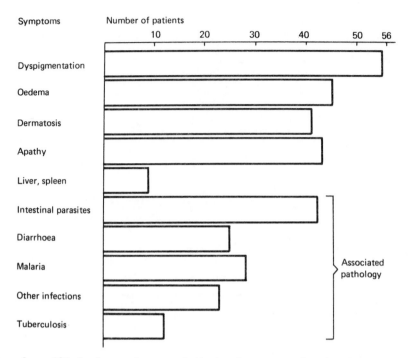

Fig. 1 : Clinical symptoms and their frequency in the 56 PEM patients investigated in this study.

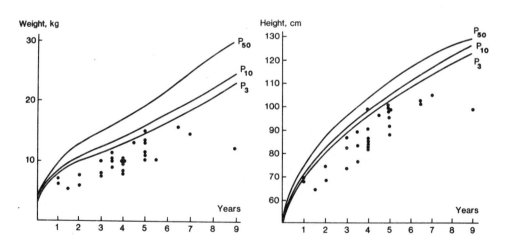

Fig. 2 : Anthropometric measurements (weight/age and height/age) of the PEM children investigated in this study.

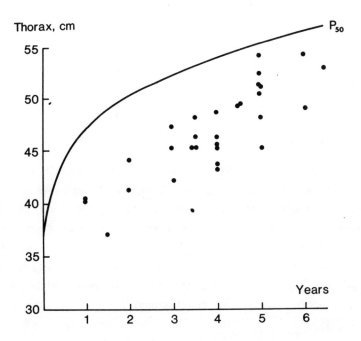

Fig. 3 : Anthropometric measurements (chest circumference/age) of the PEM children investigated in this study.

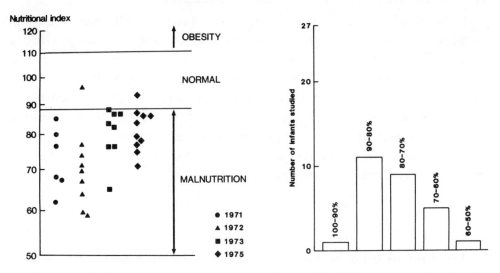

Fig. 4 : Anthropometric measurements of the PEM children investigated in this study. On the left side : nutritional index (Dugdale, 1971) and on the right side : mid-upper-arm circumference (Jelliffe, 1968).

Also the classification according to measurement independent of age shows that more than 90% of the children are well below standard. Except for two, all children fall within the malnutrition part of the nutritional index of Dugdale (1971).

Biochemical assessment

Whenever possible, the control values were taken from healthy well-fed children and adults from Yasa-Bonga, mostly hospital employees with a stable income. These controls must be clearly distinguished from the so-called rural controls, used for the amino acid study and taken from healthy but marginally nourished subsistence farmers in the villages. The results are given in the form of tables with the mean ± standard deviation. The individual values for each patient are not presented here, although they are taken into account for the general conclusions.

Serum

Protein, albumin and globulin, ceruloplasmin, prealbumin (Table III)

In children and adults, total protein is reduced but the drop in albumin is much more pronounced. The α_2- and β-globulin values are only slightly lower in the children as compared to the controls. Among adults, we find an increase in γ-globulins. The ceruloplasmin (an α_2-globulin) and pre-albumin levels are reduced in adults and children. Prealbumin, which was proposed as a useful index of PEM by Ingenbleek et al. (1972), was somewhat less reduced than albumin in these chronic cases, in which the respective half-lives of the proteins play a lesser role than the chronic impairment of the protein synthesizing machinery in the liver.

Calcium, magnesium and phosphorus (Table IV)

The great majority of our patients show a marked hypocalcemia, whereas magnesium and phosphorus are reduced only in children.

Potassium, sodium, chlorine (Table IV)

With a few exceptions, the serum levels of these electrolytes are normal in children and adults.

Total lipids and cholesterol (Table V)

All values are definitely reduced in the malnourished patients.

Haemoglobin and glucose (Table V)

Haemoglobin is extremely low in the malnourished children (6.5 versus 11-12 g % in controls) due to concomitant infections and parasites. Glycaemia tends to be reduced, especially in adults, probably because of lack of food in the sick, socially deprived women.

Immunoglobulins (Table VI)

In most patients, IgM's and IgG's are increased due to the numerous infectious episodes. IgA's are raised only in about 20% of the patients who are generally the most serious cases.

Amino acids[b] (Fig. 5)

For these determinations, blood was also drawn in the early morning, after an overnight fast. The values in the infant and adult patients are compared to a control group from the village of Mikula, the so-called rural controls. Values for European or United States (US) controls are also indicated.

(b) For the details of the results see : Antener et al.,1981a.

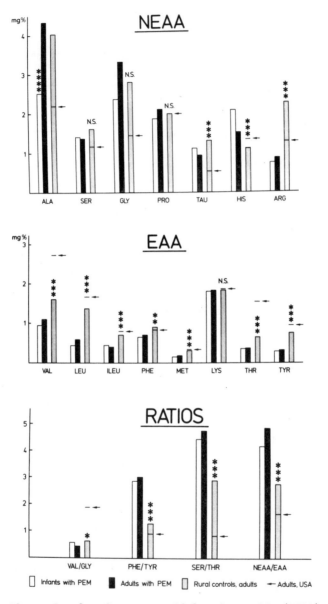

Fig. 5 : Plasma levels of non-essential amino acids (NEAA) and essential amino acids (EAA), and some NEAA/EAA ratios in PEM children, PEM adults, adult rural controls and adults Americans. N.S. : Not significant difference.

*   Different from PEM adults (p < 0.05)
**   Different from PEM children (p < 0.05)
***   Different from all PEM patients (adults and children) (p < 0.05)
****   Different from PEM adults (p < 0.01).

The serum amino acid pattern is clearly pathological and practically identical in infant and adult patients, with the exception of alanine, which is lower in infants. Actually, it is the alanine level in adults that is higher than normal. This is also found in the adult rural controls. In fact, alanine is the most glucogenic amino acid and this is a reversible reaction. The adult adapts to the high carbohydrate-low protein content of the daily diet by increasing the synthesis of alanine and possibly other non-essential amino acids. Indeed, not only alanine, but also serine and glycine are higher in the rural controls than in normal US adults. Among the other non-essential amino acids, histidine is increased in the patients due to histidase deficiency (Antener et al., 1983), serum proline is unchanged in all groups whereas the other non-essentials are decreased in the patients. Taurine is only decreased when compared to the rural controls but not to US or European standards. Indeed, the rural controls have a higher serum taurine level than US adults. This may be due to higher sulphur amino acid catabolism, because of the unsatisfactory protein status or to a reduced sulphate formation due to an unknown enzyme defect. This finding is different from that of Ghisolfi et al. (1978) in "temporary" undernutrition in European infants, in whom the lowering of the plasma taurine level was the most typical sign of malnutrition. Arginine is especially low in the patients and high in the rural controls. Actually, serum arginine is almost twice as high in the rural control as compared to US adults. This is probably due to the reduced arginase activity and represents an adaptive nitrogen-sparing mechanism due to the low protein ingestion. In the patients this mechanism is unsufficient to maintain the arginine level probably due to reduced argininosuccinase activity (Waterlow, 1968).

Like other investigators (Vis, 1975; Ghisolfi et al., 1978), we found lower values for all the essential amino acids apart from lysine. Vis (1963) and Holt et al. (1963) did not observe this particularity for lysine. The normal serum lysine value may be related to the adequate lysine content of the daily diet (Mauron, unpublished results). The threonine level is extremely low in patients and also lower in the local controls than in European or US controls. Actually, a low serum threonine value seems to be a characteristic feature of this population group. This prompted us to introduce the serine/ threonine ratio as a typical parameter for this kind of PEM. Since the threonine level in the daily diet is proportionally adequate, its low serum level must be attributed to some metabolic particularities. We found about 15 years ago a low serum threonine level in children with the Aldrich-Wisgott syndrome.

Tyrosine, a semi-essential amino acid, is also very low in PEM patients and behaves here like an essential amino acid because of the low level of phenylalanine hydroxylase (Antener et al., 1981b). The low serum tyrosine level found in the patients may partially explain their apathy and often depressed state since tyrosine is an important precursor of catecholamines and brain noradrenaline deficiency has been related to some types of depression (Gelenberg et al., 1980).

The different amino acid ratios show the same overall tendency that is observed in PEM but the valine/glycine ratio is less useful in this population than the serine/threonine ratio, because of the high glycine and relatively low valine level in the rural controls. The non-essential to essential amino acid ratio was calculated with the amino acids : ala + ser + gly + gln + glu + asn divided by val + leu + tyr + thr + ileu + phe + meth.

A very interesting feature was the high amino acid content of the erythrocytes, as compared to the serum (Table VII). A high E/S ratio for amino acids bears witness to the severity of the pathological condition since it is also found in malignancy, acute inflammation hepatitis and anaemia (Björnesjö et al., 1968). The values resemble these found by Björnesjö et al. (1968) and Mikhail et al. (1973) in malnutrition.

Enzymes

Of the eight enzymes tested, only two, namely γ-glutamyl-transpeptidase (γ-GT) and cholinesterase (CHE), are always pathological (Table VIII). The high level of γ-GT in all children and most adult patients, shows that the liver parenchyme has been damaged. It appears to be a very sensitive diagnostic tool for the type of PEM we observed in Yasa-Bonga. The low level of cholinesterase shows that protein synthesis is reduced in the liver. The activity of this enzyme also appears to be a good indicator of the severity of malnutrition.

Transaminases and the other enzymes are much less characteristic. Their level is raised only in some patients. There is a tendency towards more pathological values in infants as compared to adults. This is especially true of GPT.

γ-glutamyl-transpeptidase isoenzyme[c] (Table IX) : 6 patients with high γ-GT activity levels were investigated for the presence of iso-enzymes. Whereas in the controls, more than 90% of the activity is in the α_2-γ-GT band and the rest of the activity in the α_1-γ-GT band, all patients, except one, show a decrease in the α_2-γ-GT band and an increase of the α_1-γ-GT level. In addition, 4 patients have a band at the start (O-γ-GT) and one patient has a band in the β_1 area. These increases in the α_1 band at the origin and in the β_1 area of the protein marker are well-known pathological signs, not yet described in PEM, however.

Urine

Nitrogenous substances (Table X)

Nitrogen, urea, creatinine and hydroxyproline excretion is very low in the patients. Total nitrogen, and especially urea excretion, is a function of protein intake at the moment of admission to the hospital. The lower values in children would indicate that they had a lower protein supply than the adult women. The low creatinine excretion is a measure of the low muscle mass. It should be noted that even the adult controls had a low creatinine excretion indicating relatively low body and muscle mass in the general adult population. The creatinine/height index (Viteri et al., 1970) was calculated for the infants and found to be well below normal. Hydroxyproline excretion in infants was extremely low and so was the hydroxyproline index

---

(c) See also : Antener et al., 1980.

(Whitehead, 1965), showing that growth had practically stopped at the time of admission. It is interesting to note that in most malnourished adults, hydroxyproline excretion is also low confirming a reduced collagen turnover.

Electrolytes (Table XI)

Urinary calcium is very low in children and low in adults, but it should be noted that the controls also have relatively low calciuria. This is in relation with a relatively low calcium intake in the region, so that tubular reabsorption of calcium is almost complete. The magnesium excretion is also low in children (Harris et al., 1971) but somewhat less than the calciuria. In adults, magnesiuria is only slightly reduced in some patients. The fact that in the adult controls and patients, magnesium excretion is about 3 times higher than calciuria is worth mentioning. It is probably a reflexion of the nutrition situation in the region, since the magnesium supply is normal whereas calcium intake is at the inferior limit. The excretion of phosphate is very low in infants but much less so in adults. Since in the average diet of the population phosphorus intake is just about normal, the reduced phosphorus excretion in the sick infants is in relation to their low protein intake. Potassium, sodium and chlorine excretion is reduced in children and adults but low values are more frequent in children. The reduced potassium excretion is probably in relation with the diminished muscle mass and reduced potassium content of the tissues (Ingenbleek & Satgé, 1968). We could not verify this point since we could not make biopsis, but the low potassium excretion, in spite of normal extra-cellular potassium level in serum is probably a sign of potassium depletion of the tissues and linked to the frequent oedema observed (Golden, 1982). The reduced sodium and chlorine levels in urine are in relation with the low salt intake in the average diet.

Vitamins and trace elements

The few vitamin determinations made in urine do not allow to draw a definite conclusion. The excretion of vitamins B_1, B_2, niacine and B_6

was well below normal, showing that most patients studied were on the borderline of vitamin deficiencies.

Most of our patients, children and adults alike, had a reduced excretion of copper, zinc and manganese. The great variability of the excretion of these three trace elements in the controls should, however, be borne in mind. Since in the average diet, only copper is in short supply, the more generalized deficiency in the malnourished group can be attributed to reduced intestinal absorption.

Faeces

Some of the patients, mainly the children, had a large number of daily bowel movements and nearly all had periods of diarrhoea.

A characteristic feature of the patients at Yasa is the presence of starch (Auricchio et al., 1968), sometimes in high amounts, in the faeces, due to atrophy of the pancreas and also to the fact that about 90% of the daily diet is composed of manioc. Steatorrhoea plays a minor role when the patient enters the hospital because of the very low fat content of the daily diet (6%). At the beginning of the treatment, however, when the fat content of the diet is increased, steatorrhoea is often observed.

The determination of sugars in the ultrafiltrates of the faeces shows secondary intestinal disaccharide and monosaccharide malabsorption. Malabsorption of maltose was observed in 41% of the patients, of isomaltose in 32% and of glucose in 85%. These malabsorptions are typical signs of the atrophy of the intestinal mucosa.

DISCUSSION

Protein-energy malnutrition as described here, although it shows most biochemical signs of kwashiorkor, is of a multi-factorial type involving also energy, vitamins, minerals and trace elements. Because of the seasonal variations in the food supply, this malnutrition is recurrent with periodic relapses and increasing severity of the disease and finally resistance to dietary therapy. Discolouration is the most typical sign of PEM in that region and is probably linked to the low level of riboflavin in the average diet of the population. Hughes made a similar observation in Nigeria (1946).

An important particularity of the region of Yasa-Bonga is the high frequency of adult PEM. Our investigation shows that the adult type has the same biochemical characteristics as the generally described infantile type (Table XII). Among the different parameters measured, only two were different in infant and adult patients, namely γ-globulins and alanine. We conclude that these two forms of PEM do not represent different entities but are the expression of the same basic syndrome that does not depend on age per se but affects the vulnerable individuals whenever the food deficiency and the socio-economic deprivation falls below a critical level. In the rural environment of Yasa-Bonga, preschool children and multiparous lactating women represent the two most vulnerable groups but PEM can also be observed in old widowers.

Although it is recognized that PEM is a multi-factorial disease, the relatively high frequency of adult PEM in that region has to be linked to the very low protein level in the average diet which contains only between 4 and 5% protein-energy.

ASSESSMENT OF DIETARY REHABILITATION

Diets

Another part of our investigation was the design of dietary treatment using partially local foods[d]. Four types of diets were successively used for dietary treatment (Table XIII). There is no place to discuss here the

---

(d) Antener et al., 1977b.

rationale of these diets. They contained about 12-15% protein calories, 10-25% fat calories and the rest from carbohydrates. All diets contained locally available dried fish as main protein source, combined in some cases with lactose-free milk and caseinate. One diet (diet IV) contained only locally available foodstuffs. Complete balance sheet assays of the main nutrients (nitrogen, fat, minerals and trace elements) were made with children suffering from PEM with these diets and the results have been presented elsewhere[d]. It may be just worth mentioning that very satis-factory positive nitrogen balances could be obtained with all diets and especially with diet IV, which had a high proportion of vegetable proteins. It was most difficult to obtain positive balances for calcium and magnesium. Nutrient balance sheet assays could only be performed with infant patients so that no comparison with the adults is possible.

Biochemical assessment[e]

Biochemical parameters were measured two weeks and two months after the beginning of the treatment with the balanced diets. Moreover, some patients were studied at one year intervals for several years after they had returned to their rural homes.

Serum proteins (Fig. 6)

Transferrin is the most sensitive parameter to react to the dietary rehabilitation, followed by albumin and ceruloplasmin. Prealbumin is very slow to rise, even after 2 months of dietary rehabilitation. All four para-meters have the tendency to rise more slowly in adults. This is especially true for ceruloplasmin and prealbumin. It should be noted that in one child, all four parameters continued to fall during treatment; this child died shortly afterwards.

The behaviour of prealbumin observed here is in contrast with the findings of Ingenbleek et al. (1975) in primary malnutrition and by

---

(e) Antener et al., 1978.

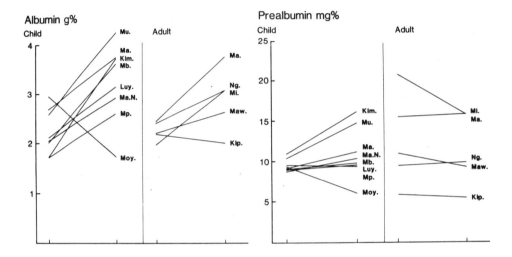

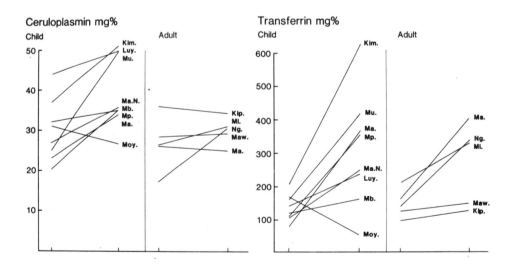

Fig. 6 : Evolution of serum albumin, prealbumin, ceruloplasmin, and transferrin during two months of dietary treatment.

Carpentier et al. (1982) in hospitalized patients. In both cases, the concentration of prealbumin reverted to normal values upon adequate refeeding. The long-lasting impairment of prealbumin synthesis we observed, in spite of adequate refeeding, may be the expression of liver damage, affecting more severely prealbumin synthesis than that of the other visceral proteins measured. The fact that prealbumin synthesis is more impaired in adult patients than in children would point in the same direction since the liver is more injured in adults because they have experienced many more episodes of malnutrition during their life. Another, less likely, explanation would be the persistance of a stressful situation during the whole period of rehabilitation, favouring the preferential synthesis of acute phase reactants (inflammatory markers) at the expense of prealbumin. This is, however, not very likely since this adaptive mechanism was shown to be operative only during the few days following the onset of acute stress (Ingenbleek, 1982).

Serum amino acids

The serum amino acid levels are rather slow to respond to dietary rehabilitation but the tendency of the essential amino acids to rise and for the non-essentials to fall, was already distinguishable after 2 weeks' treatment but the normalisation was much slower and less pronounced than for certain serum protein markers and enzymes.

Enzymes (Fig. 7)

The most interesting parameters to follow the evolution of the illness during treatment are the two enzymes : γ-glutamyl-transpeptidase (γ-GT) and cholinesterase (CHE). Concomitant decrease of γ-GT and increase of CHE to normal levels signify good response to the treatment and favourable evolution of the illness. In case the normalisation of the two enzyme levels, especially that of CHE, cannot be obtained, the prognosis is very bad indeed. Another important prognostic tool is to follow the evolution of the γ-GT isoenzymes. Diminishing of the α_1 band and disappearance of the β and 0-bands is a sign of favourable evolution.

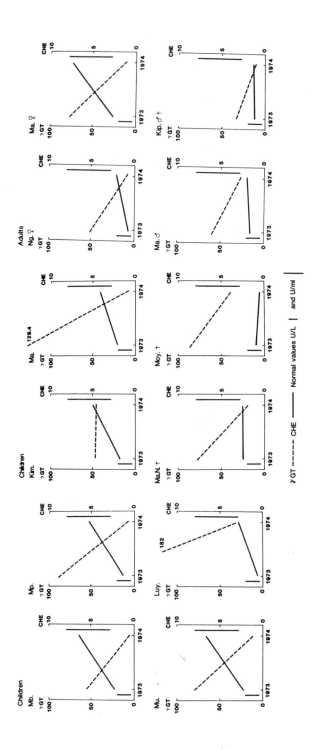

Fig. 7 : Evolution of enzymatic activity of γ-glutamyl-transpeptidase (γ-GT) and of cholinesterase (CHE) during 2 months of dietary treatment.

Urine

The excretion of urea and hydroxyproline increases markedly during treatment whereas the augmentation of creatinine excretion remains generally irregular, showing that the muscle mass recuperation is far from being uniform.

Faeces

Starch disappears progressively from the faeces upon dietary rehabilitation, especially when manioc is replaced by rice. Mono- and disaccharides do not disappear completely but their amount decreases upon treatment, showing the improvement of intestinal absorption.

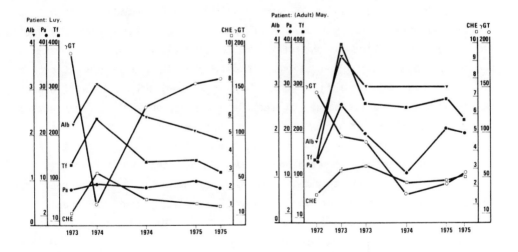

Fig. 8 : Evolution of albumin, prealbumin and transferrin serum concentrations and of the activity of γ-GT and CHE during dietary treatment and after return to the rural environment. The first result was obtained before treatment, the second one after treatment and the following ones at examinations in the rural environment.

Evolution of illness[f] (Fig. 8)

Some patients were followed over several years and a series of biochemical parameters were measured. A few examples are shown : the results
obtained with two children and two adult women during a three year observation. They show the good response of the main parameters to dietary
treatment and the slow degradation of the situation upon return to the rural
conditions, interrupted sometimes by a new improvement, due to a temporary
better food supply. It should be mentioned that not all patients experienced
these ups and downs with periodic relapses upon return to their rural
environment, some recovered patients managed to maintain a very satisfactory
nutritional status, characterized by a normalisation of most of the biochemical parameters.

Classification of parameters (Table XIV)

In order to get a clearer picture of the value of the different biochemical parameters in assessing nutritional recovery, we made an attempt to
distinguish between parameters responding more to the adequacy of the diet
and those more related to the state of the individual and having thus a
prognostic value for the evolution of the illness. We are aware that such a
distinction is very artificial, since the improvement of health will depend
on the diet and the efficiency of the dietary treatment will closely depend
on the status of the individual. However, even so, some parameters are more
directly linked with the diet and others with the status of the patient.
Thus, in the type of PEM studied in Yasa-Bonga, transferrin and albumin
react rapidly to diet therapy, whereas prealbumin is slow to respond,
depending on the status of the individual. However, when the prealbumin
value eventually increases, it is a valid sign of a recovery. Another
example is γ-GT. The drop of this enzyme's activity is a sign of the
improvement of the liver function and is thus more a measure of the status
of the patient, although in favourable cases, it reacts relatively rapidly
to the diet also. That is what is called "indirect effect" in Table XIV,

---

(f) Antener et al., 1978.

meaning that the normalisation of the γ-GT activity following dietary rehabilitation is an indirect effect of the diet mediated by the improvement of the liver parenchyme. We have tabulated in this way the different biochemical parameters, distinguishing between direct and indirect effects as regards efficiency of the diet or evolution of the illness. The table should be self-evident and is only meant to clarify the biochemical assessment of this particular type of PEM and to be a help in controlling the efficiency of the dietary management of the disease. It is also clear that such a tabulation always represents an oversimplification and that nothing can replace the close continuous observation of the individual cases who will all manifest some pecularities.

CONCLUSION

The assessment of dietary rehabilitation with the biochemical parameters chosen here shows that PEM in Yasa-Bonga is often of the chronic type, involving progressing liver damage. This is confirmed by the large number of patients treated in the same hospital for chronic liver diseases such as cirrhosis and even primary liver cancer in young adults.

To assess recovery, it is important to choose sensitive and specific methods. For the particular PEM described here, the determination of γ-GT, CHE and prealbumin appeared to give the most sensitive parameters for evaluation of the evolution of the illness, whereas transferrin was the parameter reacting most rapidly to the dietary therapy. Here again, there is no fundamental difference between infant and adult patients. However, the rapidity of the normalisation of the different parameters and hence of recovery, was generally slower in adults than in infants. This does not mean that the PEM syndrome is different in each group, but that the adult patients, having had more episodes of malnutrition during their life, manifest greater liver damage.

We conclude that adult PEM described here for the first time in some detail, including numerous biochemical parameters, is the same syndrome as that observed more generally in infants. The particular conditions in Central Kwilu with an extremely low protein intake of the general population, combined with other marginal deficiencies in vitamins and trace elements, are responsible for the relatively high frequency of adult PEM. It is thus clear that it is the general environment which imprints upon the syndrome the particular aspects observed here. The evolution of PEM in Yasa-Bonga also shows that the periodical dietary rehabilitation in the hospital is not a solution to the problem but that only a general improvement of the food situation with the introduction of crops containing more protein, such as groundnuts, Bambara groundnuts, soja, beans, corn, etc. could definitely improve the present dismal situation.

REFERENCES

Antener, I., Verwilghen, A.M., van Geert, C. & Mauron, J. (1977a) Biochemical study of malnutrition - situation before treatment. Part I : Acta Tropica 34:331-354.

Antener, I., Verwilghen, A.M., van Geert, C. & Mauron, J. (1977b) Dietary treatment of malnutrition. Balance-sheets of the main nutrients. Part II : Int. J. Vitam. Nutr. Res. 47:167-199.

Antener, I., Verwilghen, A.M., van Geert, C. & Mauron, J. (1978) Study on malnutrition. Part III : Biochemical assessment of the dietary treatment and evolution of the illness. Helv. Paediat. Acta 33:543-562.

Antener, I., Givel, F., Mauron, J. & Verwilghen, A.M. (1980) Measurement of serum iso-enzymes of glutamyl-transpeptidase in malnutrition. Acta Paediatr. Belg. 33:9-15.

Antener, I., Tonney, G., Verwilghen, A.M. & Mauron, J. (1981a) Biochemical study of malnutrition. Part IV : Determination of amino acids in the serum, erythrocytes, urine and stool ultrafiltrates. Int. J. Vitam. Nutr. Res. 51:64-78.

Antener, I., Verwilghen, A.M., van Geert, C. & Mauron, J. (1981b) Biochemical study of malnutrition. Part V : Metabolism of phenylalanine and tyrosine. Int. J. Vitam. Nutr. Res. 51:297-306.

Antener, I., Verwilghen, A.M. & van Geert, C. & Mauron, J. (1983) Biochemical study of malnutrition. Part VI : Histidine and its metabolites. Int. J. Vitam. Nutr. Res. (in press).

Auricchio, S., Ciccimarra, F., Della Pietra, D. & Moanro, L. (1968) Physio-pathologie der Stärkeverdauung beim Säugling. Akademie-Verlag, Berlin. Ernährungsforschg. 13:89-100.

Baker, E.M., Canham, J.E., Nunes, W.T., Sauberlich, H.E. & McDowell, M.E. (1964) Vitamin B_6 requirement for adult men. Am. J. Clin. Nutr. 15:59-66.

Björnesjö, K.B., Mellander, O. & Jagenburg, O.R. (1968) The distribution of amino acids between plasma and red cells in protein malnutrition. In : Calorie Deficiencies and Protein Deficiencies (Ed. McCance, R.A., Widdowson, E.M.). J. and A. Churchill, London, p. 135.

Carpentier, Y.A., Barthel, J. & Bruyns, J. (1982) Plasma protein concentration in nutritional assessment. Proc. Nutr. Soc. 41:405-417.

DeMaeyer, E.M. & Vanderborght, H. (1958) A study of different sources in the feeding of African children. J. Nutr. 65:335-352.

Dugdale, A.E. (1971) An age-independent anthropometric index of nutritional status. Am. J. Clin. Nutr. 24:174-176.

Gelenberg, A.J., Wojcik, J.D., Growdon, J.H., Sved, A.F. & Wurtman, R.J. (1980) Tyrosine for the treatment of depression. Am. J. Psychiat. 137: 622-623.

Ghisolfi, J., Charlet, P., Salvayre, N., Ser, R., Thouvenot, J.P. & Duole, C. (1978) Plasma free amino acids in normal children and in patients with proteinocaloric malnutrition : fasting and infection. Pediat. Res. 12: 912-917.

Golden, M.H.N. (1982) Protein deficiency, energy deficiency and the oedema of malnutrition. Lancet I, 1261-1265.

Harris, I. & Wilkinson, A.W. (1971) Magnesium depletion in children. Lancet II, 735-736.

Holt, L.E. Jr, Snyderman, S.E., Norton, P.M., Roitman, E. & Finch, J. (1963) The plasma aminogram in kwashiorkor. Lancet II, 1343-1348.

Hughes, E. (1946) Kwashiorkor and ariboflavinosis. Trans. Roy. Soc. Trop. Med. Hyg. 39:437-448.

Hume, E.M. & Krebs, H.A. (1949) Vitamin A requirements of human adults. Report of the Vitamin A Sub-Committee of the Accessory Food Factors Committee. Medical Research Council, Special Report Series No 264, London, pp. 145.

Ingenbleek, Y. (1982) Usefulness of prealbumin as nutritional indicator. In : Marker Proteins in Inflammation (Ed. Allen, R.C., Bienvenu, J., Laurent, Ph., Suskind, R.M.) Walter de Gruyter, Berlin, New York, pp. 405-414.

Ingenbleek, Y. & Satgé, P. (1968) Importance théorique et pratique des troubles du métabolisme du potassium et du magnésium dans le kwashiorkor. Bull. Soc. Méd. Afr. Noire Langue Fr., 13:895-904.

Ingenbleek, Y, de Visscher, M. & De Nayer, Ph. (1972) Measurement of prealbumin as index of protein-calorie malnutrition. Lancet II, 106-109.

Ingenbleek, Y., van den Schrieck, H.G., De Nayer, Ph. & de Visscher, M. (1975) Albumin, tranferrin and the thyroxine-binding prealbumin/retinol-binding protein (TBPA-RBP) complex in assessment of malnutrition. Clin. Chim. Acta 63:61-67.

Jelliffe, D.B. (1966) The assessment of the nutritional status of the community. WHO Monograph No 53, Genève.

Levy, H.L. & Barkin, E. (1971) Comparison of amino acid concentrations between plasma and erythrocytes. Studies in normal human subjects and those with metabolic disorders. J. Lab. Clin. Med. 78:517-523.

Mikhail, M.M., Patwardhan, V.N. & Waslien, C.I. (1973) Plasma and red blood cell amino acids of Egyptian children suffering from protein-calorie malnutrition. Amer. J. Clin. Nutr. 26:387-392.

Nelson, W.E. (1964) Growth and development in the infant and the child. In : Textbook of Pediatrics (Ed. Nelson, W.E.), W.B. Saunders Company, Philadelphia/London, pp. 14-60.

O'Brien, D., Ibott, F.A. & Rodgerson, D.O. (1968) Laboratory Manual of Pediatric Microbiochemical Techniques. Harper and Row, New York, p. 283.

Pereira, S.M. & Begum, A. (1974) The manifestation and management of severe protein-calorie malnutrition. Wrld Rev. Nutr. Diet. 19:1-50.

Vis, H.L. (1963) Aspects et Mécanismes des Hyperaminoacidurées de l'Enfance. Recherches sur le Kwashiorkor, le Rachitisme Commun et le Scorbut. Editions Arscia (Bruxelles), Librairie Maloine (Paris), pp. 325.

Vis, H.L. (1975) Acides aminés et Kwashiorkor. XXIVe Congrès de l'Association des Pédiatres de Langue Française. Paris, Vol. II, p. 219.

Viteri, F.E. & Alvarado, J. (1970) The creatinine height index : its use in the estimation of the degree of protein depletion and repletion in protein calorie malnourished children. Pediatrics 46:696-706.

Waterlow, J.C. (1968) The adaptation of protein metabolism to low protein intakes. In : Calorie Deficiencies and Protein Deficiencies (Ed. McCance, R.A., Widdowson, E.M.) J. & A. Churchill Ltd. London, pp. 61-73.

Whitehead, R.G. (1965) Hydroxyproline creatinine ratio as an index of nutritional status and rate of growth. The Lancet, II, 567-570.

Table I : Nutrition survey of the Kumbimashi family.
Daily nutrient intake per person.

Nutrient	Daily intake per person	Requirements[1] Moderate activity	Requirements[1] Heavy activity	Requirements met
Kcalories	2040	1930	2134	OK

Proteins	Daily intake per person	Requirements[1] min.	Requirements[1] safe allowance	Allowance met
	20 g	23 g	30 g	---

Vitamins	Daily intake per person	Dietary allowance[2] min.	Dietary allowance[2] recommendation	Allowance met
A (ß-carotene)[3]	1100 I.U.	965 I.U.[4]	3400 I.U.	--
B_1	0.53 mg		0.7 mg	--
B_2	0.45 g		1.2 mg	---
Niacin	7.3 mg		8.0 mg	OK
B_6	1.0 mg		0.77 mg[5]	OK
B_{12}	1.8 mg		2.0 mg	OK

(1) Average based on weight and age of the persons composing the family (2 men, 2 women, 1 child) using the values of the FAO/WHO Report 1973 for energy and protein requirements.

(2) Average based on the calorie intake or the weight of the individuals using the Dietary Standard of Canada 1963 (Canadian Council on Nutrition).

(3) 1 µg ß-carotene = 1 I.U. vit. A activity.

(4) Average based on minimal requirements of 25 I.U./kg for adults (Hume & Krebs, 1949) and 50 I.U./kg for the child.

(5) Average based on the values of Baker et al., 1964, for a low protein diet, corrected for body weight or age.

Table II : Nutrition survey of the Kumbimashi family.
Meeting nutrient requirements.

Mineral salts	Daily intake per person	Dietary[1] allowance	Allowance covered
Na	(0.112 + 1.96 g) = 2.07 g	?	OK
K	2.23 g	1.2 - 3.7 g	OK
Ca	0.543 g	0.54^2 - 0.8^1	±
Mg	276 mg	274 mg	OK
P	0.784 g	0.540 g	OK
Mn	4.9 mg	2.5 mg	OK
Fe	14.45 mg	13.2 mg	OK
Cu	0.83 mg	2.0 mg	--
Cr	660 mcg	250 mcg	+++
Zn	10.14 mg	15 mg	±
F	4.24 mg	1.5 mg (anticaries effect)	+++

(1) Average based on the sex and the age (or weight) of the persons composing the family, using the Recommended Dietary Allowances 1980 of the Nutritional Academy of Sciences, Washington DC.

(2) Based on the Dietary Standard of Canada 1963 (Canadian Council of Nutrition).

Table III : Serum proteins (Mean ± S.D.)

	Units	Infants		Adults	
		Patients	Controls	Patients	Controls
Total protein	g/l	54.6 ± 8.7	71.1 ± 6.9	61.7 ± 10.3	76.1 ± 6.6
Albumin	g %	1.89 ± 0.53	3.31 ± 0.35	1.93 ± 0.50	4.36 ± 0.38
α_1-Globulin	g %	0.27 ± 0.09	0.32 ± 0.07	0.31 ± 0.06	0.27 ± 0.08
α_2-Globulin	g %	0.56 ± 0.13	0.71 ± 0.18	0.57 ± 0.13	0.48 ± 0.10
β-Globulin	g %	0.59 ± 0.15	0.75 ± 0.15	0.66 ± 0.18	0.76 ± 0.13
γ-Globulin	g %	2.08 ± 0.51	2.04 ± 0.61	2.69 ± 0.67	1.80 ± 0.49
Ceruloplasmin	mg %	31.1 ± 8.7	56.1 ± 15.3	28.9 ± 7.4	42.4 ± 12.9
Prealbumin	mg %	10.2 ± 4.15	14.2 ± 2.1	11.5 ± 5.7	20.9 ± 7.6

Table IV : Serum electrolytes (Mean ± S.D.)

	Units	Infants		Adults	
		Patients	Controls	Patients	Controls
Potassium	mEq/l	4.5 ± 0.49	4.52 ± 0.59	4.96 ± 0.51	4.63 ± 0.29
Sodium	mEq/l	136.2 ± 5.80	138.0 ± 1.94	139.6 ± 3.7	140.6 ± 2.3
Chlorine	mEq/l	103.6 ± 7.20	103.2 ± 2.97	104.9 ± 4.5	102.9 ± 3.3
Calcium	mg %	8.3 ± 0.80	9.1 ± 0.45	8.5 ± 0.36	9.8 ± 0.28
Magnesium	mg %	1.83 ± 0.25	2.22 ± 0.11	1.92 ± 0.24	1.99 ± 0.34
Phosphorus	mg %	3.86 ± 0.75	5.20 ± 0.67	3.79 ± 0.8	3.98 ± 0.32

Table V : Serum glucose and lipids (Mean ± S.D.)

| | Units | Infants | | Adults | |
		Patients	Controls	Patients	Controls
Glucose	mg %	63.5 ± 13.6	85 ± 15.0	46.7 ± 4.9	85 ± 15.0
Total Lipids	g/l	4.93 ± 1.83	6.89 ± 2.68	4.94 ± 1.37	5.84 ± 1.23
Cholesterol	mg %	89.9 ± 28.6	132.6 ± 24.8	99.6 ± 38.3	165.0 ± 38.6

Table VI : Serum immunoglobulins (Mean ± S.D.)

	Units	Infants		Adults	
		Patients	Controls* (Range)	Patients	Controls* (Range)
Ig A	IU/ml	105.9 ± 33.3	(30 - 100)	165.6 ± 88.6	(40 - 190)
Ig G	IU/ml	291.8 ± 85.2	(60 - 125)	361.1 ± 71.9	(70 - 160)
Ig M	IU/ml	328.4 ± 120.7	(60 - 200)	645.6 ± 579.6	(110 - 370)

(*) O'Brien et al. (1968)

Table VII : Ratio of amino acid levels in erythrocytes and serum
(9 children)

Amino acids	Erythrocytes mg % + SD (E)	Serum mg % + SD (S)	t test between E and S	Erythrocytes/ Serum	E/S normal values[1]
Valine	5.27 ± 0.90	1.30 ± 0.12	P < 1%	4.05	0.82
Proline	2.08 ± 0.49	2.12 ± 0.41	NS	0.98	--
Alanine	12.30 ± 1.17	2.88 ± 0.44	P < 0.1%	4.27	2.15
Lysine	4.80 ± 0.97	1.24 ± 0.15	P < 1%	3.87	0.87*
Leucine	5.23 ± 1.23	0.32 ± 0.09	P < 1%	16.34	1.2
Tyrosine	1.80 ± 0.35	0.24 ± 0.05	P < 1%	7.5	1.3
Threonine	1.99 ± 0.34	0.35 ± 0.05	P < 1%	5.69	1.50
Serine	5.53 ± 1.03	1.49 ± 0.14	P < 1%	3.71	2.85
Glycine	8.72 ± 1.51	2.23 ± 0.20	P < 1%	3.91	3.05
Histidine	2.03 ± 0.35	1.60 ± 0.24	NS	1.27	0.98*
Taurine	3.77 ± 0.46	1.02 ± 0.17	P < 0.1%	3.70	0.82
Isoleucine	1.29 ± 0.26	0.57 ± 0.08	P < 5%	2.26	0.88
Arginine	1.83 ± 0.62	0.63 ± 0.15	NS	2.90	0.45*
Phenylalanine	2.62 ± 0.48	0.53 ± 0.05	P < 1%	4.94	1.18
Ornithine	1.89 ± 0.31	0.54 ± 0.08	P < 1%	3.50	1.15*
Methionine	1.24 ± 0.23	0.17 ± 0.03	P < 1%	7.29	1.02
Glutamine	1.02 ± 0.23	2.07 ± 0.17	P < 1%	0.49	2.0
Glutamic acid	5.25 ± 1.16	3.31 ± 0.52	NS	1.59	3.93*
Asparagine	1.01 ± 0.38	1.11 ± 0.21	NS	0.91	--
Phenyl/Ty.	1.46 ± 0.07	2.81 ± 0.47	P < 5%	0.52	0.91
Valine/Gly.	0.63 ± 0.07	0.61 ± 0.06	NS	1.03	0.27
Ser./Thre.	2.71 ± 0.16	4.78 ± 0.72	P < 5%	0.57	1.9
NEAA/EAA	1.88 ± 0.13	3.92 ± 0.34	P < 0.1%	0.48	1.27

(1)   according to Björnesjö et al. (1963)

(*)   according to Levy et al. (1971).

Table VIII : Serum Glutamyltranspeptidase (γ-GT) and Cholinesterase (CHE)
(Mean ± S.D.)

| | Units | Infants | | Adults | |
		Patients	Controls	Patients	Controls
γ-GT	U/l	77.2 ± 45.8	8.82 ± 4.12	55.4 ± 35.3	8.82 ± 4.12
CHE	U/ml	1.54 ± 0.66	(3 - 8)*	1.96 ± 1.05	(3 - 8)*

(*) Range

Table IX : Serum γ-glutamyltranspeptidase isoenzymes

Controls + Patients	γ-GT U/l	α_1-γGT %	α_1-γGT U/l	α_2-γGT %	α_2-γGT U/l	β_1-γGT %	β_1-γGT U/l	O-GT %	O-GT U/l
Control 1	4	9	0.3	91	3.7	--	--	--	--
Mb.	57	--	--	100	57	--	--	--	--
Mp.	86	72	61.9	28	24.1	--	--	--	--
Moy.	86	62	53.3	36	31	--	--	2	1.7
Kim.	47	52	24.5	38	18.8	--	--	10	4.7
Mas.	129	72	92.9	27	34.8	--	--	1	1.3
Luy.	182	22	40	38	69.2	10	18.2	30	54.6
Control 2	9	8	0.7	92	8.3	--	--	--	--

Table X : Urinary N-metabolites (Mean ± S.D.)

	Units	Infants		Adults	
		Patients	Controls	Patients	Controls
Nitrogen	g/24 h	0.4 ± 0.2	1.3 - 2.2[+]	1.2 ± 0.6	1 - 4[+]
Urea	g/24 h	0.5 ± 0.3	1.4 - 4.1[+]	1.8 ± 1.3	2 - 3[+]
Creatinine	mg/24 h	99.8 ± 53.6	186 - 311[+]	285.7 ± 104.2	300 - 953[+]
Hydroxyproline	mg/24 h	6.9 ± 3.1	38.9 ± 8.7[*]	14.5 ± 7.8	20.8 ± 7.72
			62 ± 14.2[**]		
Hydroxyproline	Index	0.7 ± 0.29	3.0 ± 0.8[°]	---	---
Creatinine height	Index	0.45 ± 0.19	0.9 ± 1.2[∞]	---	---

(*)   aged 1-4 years
(**) aged 4-8 years
(+)   Range
(°)   Whitehead (1965)
(∞)   Viteri et al. (1970).

Table XI : Urinary electrolytes (Mean ± S.D.)

	Units	Infants		Adults	
		Patients	Controls (Range)	Patients	Controls (Range)
Potassium	mEq/24 h	10.5 ± 8.8	19 - 53	16.4 ± 12.4	29 - 60
Sodium	mEq/24 h	11.9 ± 9.8	12 - 27	21.9 ± 19.9	33 - 158
Chlorine	mEq/24 h	14.6 ± 9.2	22 - 42	25.0 ± 17.8	33 - 159
Calcium	mg/24 h	0.6 ± 0.9[+]	1.1 - 7.4[+]	10.8 ± 7.5	3 - 25
Magnesium	mg/24 h	1.2 ± 0.9[+]	0.9 - 5.2[+]	36.0 ± 25.2	30 - 90
Phosphorus	mg/24h	31.2 ± 29.1	98 - 247	102.3 ± 82.3	127 - 381
Copper	µg/24 h	15.0 ± 9.2	13 - 23	37.6 ± 19.9	40 - 60
Zinc	µg/24 h	113.5 ± 74.2	121 - 254	341.6 ± 260.6	256 - 1500
Manganese	µg/24 h	3.3 ± 2.4	3 - 10	4.2 ± 3.1	3 - 23

(+)   mg/24 h/kg

Table XII : Biochemical assessment of PEM

Parameter	Children	Adults
Total Protein	--	-
Albumin	---	---
Prealbumin	-	--
γ-Globulins	0	+
Ceruloplasmin	--	-
NEAA/EAA ratio	++	++
Phe/Tyr ratio	++	++
Ser/Thr ratio	++	++
Val/Gly ratio	-	-
Ig A	+	+
Ig G	+++	+++
Ig M	+++	+++
γ-GT	+++	+++
CHE	--	--
Urea	--	--
Creatinine	---	--
Hydroxyproline	---	--

- reduction ⎫
                ⎬ compared to controls
+ increase   ⎭

Table XIII : Composition of the diets

Diet I 1972	Diet II 1973	Diet III 1975	Diet IV 1975
Manioc	Rice	Rice	Soya and corn
Dried fish	Dried fish	Dried fish	Dried fish
AL110[1] or Hyperprotidine[2]	AL110 or Hyperprotidine	Aledin[5] or Hyperprotidine	Caterpillars
			Peanuts
LAD[3]	LAD	Maltrinex[6]	Vegetables
MCT	Arobon	Peanuts	
Arobon[4]	Fruit and vegetable	Caterpillars	
Fruit and vegetable		Fruit and vegetable	
Digestive ferments		Correction of electrolytes	
Ca-gluconate			
Vitamins		Added vitamins and oligoelements	

(1)   Lactose-free milk
(2)   Casein
(3)   Lactalbumin hydrolysate
(4)   Powdered carob meal (antidiarrheic)
(5)   Lactose-free milk
(6)   Maltodextrin (Sopharga)

Table XIV : Biochemical assessment of recovery
(children and adults)

Parameters	Efficiency of diet	Evolution of illness
Transferrin	Most sensitive	indirect
Albumin	sensitive	indirect
Prealbumin	non-sensitive	slow to respond ↑ positive
Ceruloplasmin	sensitive (children) slow response in adults	indirect
γ-GT	indirect	↑ very negative
γ-GT isoenzymes	indirect	α_1 + β negative
CHE	indirect	↑ very positive
Urea	very sensitive	indirect
Hydroxyproline	indirect	indicator for growth ↑ positive
Excretion of electrolytes	± sensitive, slow	↑ positive
Starch + sugar in faeces	indirect	↑ negative

PUBLIC HEALTH/CLINICAL SIGNIFICANCE OF INORGANIC CHEMICAL ELEMENTS

Mohamed ABDULLA

Department of Clinical Chemistry, •University Hospital, Lund, Sweden
and Institute for Community Health Sciences, Dalby, Sweden

SUMMARY

Food tables documenting the concentration of various nutrients in individual foods do not provide satisfactory information on the contribution of essential inorganic trace elements by prepared meals. Only direct analysis of the actual food consumed during a 24-hour period can give an accurate estimate of the dietary intake. Employing the duplicate portion technique, we have analysed the concentration of a number of inorganic chemical elements in 882 dietary samples from various population groups in Sweden. The intake of several elements including potassium, magnesium, zinc, copper and selenium, in the normal mixed diets is low when compared with recommended dietary allowances (RDA). Vegetarian diets are richer in these elements than ordinary non-vegetarian diets. Plasma levels of trace elements are often poor indicators of body status. A clinical follow-up of a group of pensioners in Dalby, Sweden, for a period of 10-12 years has not indicated any specific signs or symptoms of trace element deficiencies. Vulnerable groups, namely, children, pregnant women, alcoholics and the elderly, may, however, be more susceptible than the population in general to marginal deficiencies resulting from normal everyday consumption of Western diets.

* * *

Advances in analytical methodology and sophisticated instrumentation have certainly helped to bring about a breakthrough in the field of trace element research during the last two decades. Using powerful analytical techniques, such as atomic absorption spectrophotometry, neutron activation, mass spectroscopy and polarography, researchers have demonstrated the presence in living systems and their food chain of a number of inorganic chemical elements - elements that were considered up till now as contaminants or accumulated material, and as such, of no importance in human health and disease. The latest estimates indicate that most tissues of a healthy adult contain 40-60 of the basic elements represented in the periodic table (Hamilton, 1979; Abdulla et al., 1982a). In studies concerning the health and diseases of man, these are traditionally classified as essential or non-essential according to their role in vital metabolic processes. The bulk of living matter consists of the first eleven elements in the periodic table. Such so-called trace elements as chromium, cobalt, copper, iron, manganese, zinc and selenium constitute somewhat less than 1% of the total human body mass. The number that are known to be essential for life is likely to increase in the future with advances in our knowledge concerning the role of the elements in metabolic reactions.

In spite of the tremendous progress in trace-element nutrition research during the last few decades, the biological roles as well as the minimum requirements of many elements are still hypothetical. Some of these that are present in the human body in very low concentrations (ppb levels) are not thought to have any specific biological function. At the same time, it is generally believed that the minimum requirements of essential chemical elements, including chromium, copper, manganese and selenium, are so low that a pure nutritional deficiency of these rarely occurs in man. On the other hand, the adequacy of a modern all-round Western diet as regards trace elements is presently being debated. Marginal deficiencies of several trace elements have been shown to exist in parts of the populations of industrialized countries (Mertz, 1970).

From a public health point of view, it is important to assure the general population that the intake of all nutrients, including the essential elements, is adequate in the average, normal daily diet. At the same time, the ideal diet should not contain more than the permitted levels of toxic heavy metals. Except for occupational exposure, the major pathway through which the chemical elements enter the human body is via the food chain. Requirements established by controlled balance studies exist only for a few elements. Information concerning the dietary intake of essential inorganic elements from prepared meals is very limited at this time. Further, the data currently available are unsatisfactory, since most of them are based on conventional techniques, such as dietary history and interview studies in combination with food tables. Food tables have limitations when used for the assessment of trace-element intakes. They do not, for example, take into consideration the effect of preparing and cooling, prior to consumption, which often alter the composition of raw food materials. Moreover, food tables have no data for some elements, which occur in very minute amounts in basic foodstuffs.

Another important problem in trace-element nutrition is the diagnosis of deficiency. Severe deficiencies that require immediate medical attention occur only rarely in affluent countries. Such cases are reported occasionally in hospitalized patients receiving total parenteral nutrition (Jeejeebhoy et al., 1977; Mertz, 1981a). On the other hand, there are indications that marginal deficiencies of magnesium, copper, zinc and selenium, are fairly common in parts of the population of industrialized countries (Abdulla et al., 1981c; Mertz, 1981b). A lack of characteristic symptoms and diagnostic techniques is the major reason that marginal deficiencies are not identified at an early stage. Even when the dietary intake is restricted, normal body functions are maintained for a certain period of time by homeostatic mechanisms and by making use of the body reserves. An ideal approach to study the long-term effect of marginal consumption of trace elements is to follow vulnerable groups in the general population over extended periods for possible signs and symptoms of deficiency. Another way of detecting the existence of marginal deficiency rests in therapeutic trials. The response to iron supplementation in iron-deficiency anaemia is a good illustration. Unfortunately, there are no clinical parameters, such as haemoglobin levels in blood, to correlate the therapeutic response to other trace elements. Changes in the activity of metalloenzymes in blood and other tissues can sometimes be related to the concentration of trace elements. In our own studies, we could show that the activity of delta-aminolevulinic acid dehydratase in red blood cells could be activated by oral supplementation of zinc (Abdulla & Haeger-Aronsen, 1971; Abdulla & Svensson, 1979, Abdulla, 1980; 1981). Similarly, alkaline phosphatase activity in plasma and other biological fluids has been found to be correlated with zinc concentrations (Kirchgessner & Roth, 1980).

At Dalby in Southern Sweden, the Swedish Medical Board supports a unit for primary medical care. Dietary habits and the general health of the population is one of the research projects in this unit, wherein the duplicate portion technique has been standardized for studying the dietary intake of all nutrients in the diet of various population groups in Sweden (Borgström et al., 1975). One group of old age pensioners has been followed for a period of 10-12 years. I shall, in this paper, briefly discuss the technique used to study the dietary intake of essential and toxic inorganic chemical elements (Abdulla et al., 1979a) and the major findings of our follow-up studies.

MATERIALS AND METHODS

The material consisted of 882 dietary, 243 urine and 457 blood samples. The particulars of the participants are shown in Table I.

The duplicate portion technique was employed to collect the daily dietary samples. The duplicate portion was a copy collected visually while eating to represent as exactly as possible the food and fluids consumed during a 24-hour period. The subjects were briefed thoroughly by a trained dietician about the sampling procedure. On the day of sampling, a written protocol was prepared to indicate the rough quantities and types of solid foods and fluids consumed. The material collected during the day was pooled and later homogenized and 100-200 ml of the homogenized material lyophilized. This

Table I : Particulars of the participants (various population groups)
included in the Dalby study

Groups investigated	Age range	Sex M	Sex F	No of samples collected Diets	Urine	Blood
Healthy adults	25 - 60	10	10	140	-	-
Healthy women	30 - 79	-	60	360	180	60
Pensioners	67	17	20	259	-	370
Vegans	49 - 55	3	3	24	24	6
Lactovegetarians	49 - 52	3	3	24	24	6
Children	14 - 15	7	8	75	15	15

material was then extracted using chloroform and methanol to determine the fat-soluble components of the diet. The fat-free powder was then dried, quantitated and used to estimate the concentration of the inorganic elements.

The 60 women, 12 vegetarians and 15 teenagers who participated in our study also collected 24-hour urine samples corresponding to the days of food sampling. From most of the participating subjects, blood samples were drawn at least on one occasion. In the case of the pensioners, blood samples were collected on every occasion of a clinical follow-up. Zinc and magnesium levels were measured in whole blood as well as in plasma, whereas the other elements were determined only in plasma. The concentrations of nickel, chromium, iodine, lead, cadmium and mercury were also measured in the diets of the pensioners.

For analysing the inorganic elements, approximately 0.5 g of the fat-free powder was wet-ashed at below 250°C using concentrated nitric, sulphuric and perchloric acids. The details of the procedure are reported elsewhere (Borgström et al., 1975). Following wet-ashing, the samples were diluted with de-ionized water and kept in the cold room until analysis. For the analysis of calcium and magnesium, the samples were diluted with 2% w/v lanthanum chloride. Sodium and potassium were analysed by flame photometry; the others were determined by atomic absorption spectrophotometry except iodine which was quantitated by a spectrophotometric method (Abdulla et al., 1979b). A number of 24-hour dietary samples were also subjected to neutron activation for comparison of results with those obtained by other techniques.

The clinical examination in the case of pensioners consisted of a standardized questionnaire and a routine clinical examination, including the analysis of blood and urine. The laboratory data included height, weight, blood pressure, haemoglobin, leucocyte counts, erythrocyte sedimentation rate, blood folate, serum cobalamine, plasma tocopherol, serum vitamin A, serum proteins, serum triglycerides, serum cholesterol, serum T_3 and T_4, blood glucose and liver function tests. The concentration of zinc, copper, iron, magnesium, calcium and selenium was also measured in the plasma. The extensive clinical and laboratory investigations of the pensioners were undertaken at 2-year intervals for 12 years. These investigations are still being continued in those who are alive.

The material also consisted of 24 hospital diets (breakfast, lunch and dinner). These were included in order to make a comparison of the results by computation employing standard food tables, as well as by chemical analysis. These hospital dietary samples were treated in the same way as the other diets prior to chemical analysis.

Urine samples, whenever collected, were analysed for the estimation of the electrolytes sodium and potassium and nitrogen. We also measured the excretion of calcium, magnesium, zinc and copper in a number of urine samples.

RESULTS

The losses of the inorganic elements during lyophilization and fat extraction were calculated by analysing the fresh homogenates and the fat-free, lyophilized powder. In most cases, the losses were less than 5%. Recovery studies indicated that the wet-ashing procedure resulted in a 5-10%

loss of sodium, potassium and magnesium. For the remaining elements, the losses were less than 5% (Abdulla et al., 1982a). Table II shows the concentration of calcium, magnesium, sodium and potassium in the "common" (the averages of the diets of adults, women and pensioners), vegan and lactovegetarian diets. (The results of the diets from teenagers were not completely ready at the time of preparation of this manuscript). Fig. 1 shows the intake levels of iron, zinc, copper and selenium in the diets as percentages of the recommended dietary allowances levels (RDA) (National Academy of Sciences, Washington, 1980). As can be observed, the concentration of most of the elements in the common diet is low when compared with the RDA levels. This is especially so for potassium, magnesium, zinc, copper and selenium. The vegetarian diets, on the other hand, have much higher concentrations of these elements when compared with those of the common diet. The selenium content in the vegan diet is extremely low, whereas the lactovegetarian diet has the highest concentration of all of the element. The lactovegetarian diet, on the whole, is richer in most trace elements than the so-called common diet, but slightly poorer than the vegan diet except for selenium. Mercury, cadmium and lead in the pensioners' diets is below the permitted levels proposed by the FAO/WHO (Expert Committee on Food Additives, 1972); chromium, nickel and iodine intakes, on the other hand, are far more than adequate when compared with the present recommended levels (RDA, 1980).

The results obtained using food tables and by chemical analysis of the hospital diets showed that the computed data, in general, gave higher values than those obtained by chemical analysis. The food tables normally do not have data for the concentration of several trace elements such as copper, chromium and selenium, and as such are not suitable for comparative studies. In the present investigation, we only compared the concentration of iron, calcium, sodium and potassium. Fig. 2 shows the data for the concentration of iron obtained using food tables and by chemical analysis.

Table III shows the results of the analysis of plasma and urine samples for the concentration of calcium, magnesium, zinc and copper. Those for sodium, potassium, and in some cases, selenium and chromium, are not tabulated here. In general, most of the values obtained in the present investigation lie within the normal range.

Table II : Concentration (mean and range, mmol/l) of sodium, potassium,
calcium and magnesium in the common and vegetarian diets

Elements	Types of diets		
	Common	Vegan	Lactovegetarian
Sodium	102 (16-294)	96 (23-168)	104 (25-188)
Potassium	52 (15-93)	103 (51-163)	125 (53-160)
Calcium	17 (5-67)	16 (3-33)	23 (12-41)
Magnesium	9 (4-20)	22 (10-35)	18 (12-32)

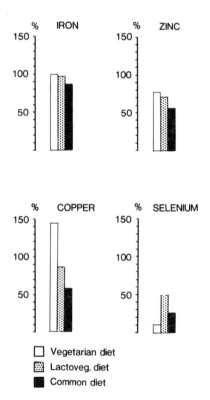

Fig. 1 : Concentrations of iron, zinc, copper and selenium in the diets
as percentages of the RDA levels.

Table III : Concentration of calcium, magnesium, copper and zinc
in plasma and urine (mean and SD, mg/1)

Groups		Calcium	Magnesium	Copper	Zinc
Lactoveg.	plasma	92.2 ± 6.7	17.2 ± 1.9	0.950 ± 0.170	0.91 ± 0.16
	urine	99.5 ± 60.9	67.1 ± 20.1	0.038 ± 0.015	0.37 ± 0.24
Vegans	plasma	---	23.1 ± 1.2	1.090 ± 0.240	1.14 ± 0.31
	urine	56.9 ± 18.1	80.5 ± 24.8	0.040 ± 0.021	0.32 ± 0.11
Pensioners	plasma	94.5 ± 9.7	18.4 ± 1.7	1.280 ± 0.370	0.80 ± 0.12
	urine	---	---	---	---
Children 15 ys-old	plasma	133.0 ± 5.7	19.1 ± 1.3	1.000 ± 0.130	1.15 ± 0.11
	urine	---	---	---	---

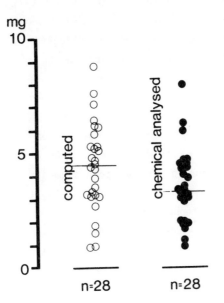

Fig. 2 : Comparison of the concentration of iron computed using food
tables and that obtained by chemical analysis (hospital diets).

DISCUSSION

In the past, food tables documenting the composition of individual food items have been extensively used for mass surveillance of food quality, especially with regards to the macronutrients proteins, fat and carbohydrate. The tables are generally based on the analysis of raw foodstuffs and, as such, may not be accurate for estimating the intake of micronutrients, especially the trace elements, from prepared meals ready for consumption. Besides, they do not provide values for chromium, selenium, manganese and arsenic. Development of the duplicate portion technique has eliminated many such problems. One major advantage of the technique is the possibility to store dietary samples for an unlimited period of time thus permitting later analysis for any particular element of interest that has been omitted at the time of the first investigation. This is pertinent because more and more elements in the periodic table are gaining importance in the field of human nutrition.

Because the results of the chemical analysis generally show lower values than those computed using food tables, the duplicate portion technique has been criticized in that it may not represent what people really consume on a day-to-day basis. This aspect has been investigated in detail in the study of the 60 women (Thulin et al., 1980; Abdulla et al., 1981b) in whom the urinary excretion of electrolytes and nitrogen was measured on two different occasions without the simultaneous collection of dietary samples. On a third occasion after an interval of two years from the first, these women collected six daily portions of food and three 24-hour urine samples corresponding to the days of food collection. Once again, we analysed the intake levels and excretion of sodium, potassium and nitrogen. There was no significant difference between the three investigations. Thus, it appears that the food collection technique did not influence the pattern of normal eating habits during the period of our investigations. Analysis of hospital diets by computation using food tables and by chemical analysis also show that the latter approach gives lower values than the computed data. A close look at the available literature indicates that the previous estimate of trace-element intakes was based on rough estimates from the levels found in the raw food materials. Our results from different series of dietary samples are reproducible and in agreement with the results of others who employed the same technique (Smith et al., 1982).

Our most important finding is the low content of several essential elements in the so-called common diet as compared with the recommended dietary allowance (RDA) levels. Similar results have been found in the diets of the general population in other affluent countries (Klevay et al., 1979; Smith et al., 1982). Overconsumption of highly refined food items may partly contribute to the low intake level of some of the elements, such as zinc, copper and selenium (Underwood, 1977). The vegetarian diets consisting mainly of unrefined foodstuffs, indeed, have a higher content of most essential trace elements; however, they may not be biologically fully available for absorption. High-fibre content in the diet can reduce the absorption of several trace elements (Sandstead et al., 1976) and vegetarian diets are fairly rich in fibre (Abdulla et al., 1981a). More detailed and long-term studies are necessary to evaluate whether vegetarian diets are a better or an inferior source of bioavailable trace elements than the normal mixed diets. Such studies may be conducted in populations such as the Hindu Brahmins in India who, because of religious beliefs, have been vegetarians for generations.

Although the plasma levels of the elements investigated are found to be in the normal range, it is difficult to interpret the results with regard to the body status of these elements. At present, no single laboratory parameter indicates the body status of most of the inorganic elements. Traditionally, the concentrations are measured in plasma or serum, for it is a simple, fast and inexpensive method. Plasma or serum levels, however, may not be good indicators for many, including zinc, copper, magnesium and iron, as most of these are intracellular (Abdulla, 1982). Moreover, a low or high level of an element in the plasma or serum may not always indicate a dietary deficiency or excess of the element in question. Very often, changes in plasma levels are due to a shift of the element from the transport media to other body compartments (Abdulla, 1982). A wide range of conditions, such as acute and chronic infections, inflammatory diseases, burns and myocardial infarctions, are known to influence the plasma levels of trace elements (Beisel et al., 1976; Powanda, 1982). A number of pharmaceutical drugs that are in common use are also known to influence the plasma levels of several trace elements (Weismann, 1980).

For reliable information on the body status of trace elements, direct measurements in vivo - whole-body counting using radioisotopes and analysis of vital biopsy materials : viz. liver, kidney, muscle and bone sample - are indicated. For routine purposes, these methods are, however, not practicable. So much so, that until simple or sensitive methods for the assessment of trace-element status are developed, dependence will be on indirect measurements of plasma, urine, hair and nail samples. To assess the body status of trace elements, it is advantageous sometimes to combine the analysis of urine and plasma samples. Analysis of nucleated tissue such as leucocytes has recently been found to be a reliable indicator of body status of trace elements including zinc when compared with that of plasma levels alone (Whitehouse et al., 1982). Because the turn-over rate of trace elements in leucocytes is fast, deficiency states may be reflected at an early stage, if the cellular concentration of these elements is accurately measured. Specific enzyme studies may be another possibility for assessment of trace elements status in the future.

As pointed out earlier in this paper, the right approach to detect a deficiency of trace elements is to follow clinically the vulnerable groups in the general population, namely, children, pregnant women, alcoholics and the elderly, for prolonged periods of time. This may not be possible in many countries due to economic and social problems. In the present study, we have followed the health status of the pensioners in Dalby, Sweden, for 10-12 years. Although the elderly people of industrialized countries constitute almost 25% of the total population, this group may not be the ideal one to follow with regards to trace element deficiencies. With advancing age, a gradual deterioration of many physiological functions are to be expected. Energy requirements decline gradually during the last decades of life. Deficiencies in the dental status and digestive problems certainly influence the dietary habits of the elderly. Moreover, elderly people are more often subjected to various drug therapies and this may influence trace-element metabolism (Abdulla et al., 1977).

Of the 37 Dalby pensioners clinically followed during the last 10-12 years, 27 are still alive today. From the clinical evaluations, which always included an elaborate laboratory data, we are unable to single out any signs or symptoms that can be attributed to the low intake of the elements we investigated.

This brings up the very important question concerning the adequacy of the present RDA values. The RDA values are subject to revision from time to time and it is often difficult to draw conclusions regarding the consequences of low intake levels of many essential elements. Moreover, there are no internationally acceptable RDA levels for many trace elements. Growing children and/or pregnant women are a more ideal group to study the effect of marginal intakes of trace elements. These groups are more susceptible to marginal deficiencies of trace elements, such as iron, zinc and copper, than the rest of the population (Mertz, 1970; Hambidge et al., 1972; Jameson, 1976; Abdulla et al., 1982b).

Another approach in the diagnosis of marginal deficiencies of trace elements rests in therapeutic trials. Many of the essential elements, for example, zinc, copper, iron and selenium, are non-toxic in therapeutic doses. Studies in Egypt and Iran on zinc deficiency and in China on selenium deficiency are good examples illustrating the positive effect of therapeutic trials (Prasad, 1976; Chen et al., 1980). In pregnant women with very low plasma zinc levels, oral zinc supplement had beneficial effects when compared with a control population (Jameson, 1976). The effect of zinc during pregnancy in humans is very similar to that which has been observed in experimental animals (Hurley & Swenerton, 1966; Mathur, 1978). Safe enrichment of table salt with iodine and the dramatic response of nutritional goitre is a classic example of therapeutic nutrition. Similarly, the response of iron-deficiency anaemia to iron supplementation has been recognized for centuries. It might be worthwhile promoting similar strategies of fortification and supplementation programmes to deal with deficiencies of other trace elements such as zinc, copper, manganese and selenium. During the last few decades, more attention has been directed to studies in animals than in man to elucidate the health effects and interactions of trace elements. It is perhaps more auspicious now to concern ourselves with human health problems in this context.

CONCLUSIONS

The duplicate-portion food sampling technique is a simple and reliable procedure to study the dietary intake of essential and toxic trace elements. Collection of 24-hour urine samples may provide additional information concerning the adequacy of the sampling. Chemical analysis of the dietary intake of the population groups investigated gave lower values than are obtained by computations using food composition tables; hence, their use may not be ideal for estimating the dietary intake of trace elements from prepared meals.

Chemical analysis indicated lower concentrations of the elements potassium, magnesium, zinc, copper and selenium in the ingested diets when compared with the RDA values. Vegetarian diets, however, had higher concentrations of these elements than non-vegetarian diets; the relatively high content of raw food materials in the vegetarian diets may be one reason. Nevertheless, it is not clear whether these elements are biologically available.

At present, there are no suitable techniques for the early diagnosis of marginal deficiencies of trace elements. Plasma levels are poor indicators of body status. A clinical follow-up study of elderly pensioners at the Dalby Health Center, Sweden, for a period of 10-12 years did not reveal any specific signs or symptoms of a deficiency of any trace element. It is imperative to direct research to developing simple and sensitive techniques for the assessment of trace-element status in vulnerable groups, such as children, pregnant women, alcoholics and the elderly.

REFERENCES

Abdulla, M. (1980) Effect of oral zinc intake on the biological effect of lead. In : Mechanism of Toxicity and Hazard Evaluation. Proceedings, Second International Congress of Toxicology (Ed. B. Holmstedt, R. Lauwerys, M. Mercier & M. Roberfroid) Elsevier, North Holland Biomedical Press, pp. 599-602.

Abdulla, M. (1981) Delta-aminolevulinic acid dehydratase activity in erythrocytes as a test in lead exposure. Publication from the Unit for Community Care Sciences. National Board of Health and Social Welfare, Dalby, Sweden, pp. 1-23.

Abdulla, M. (1982) How inadequate is the plasma zinc as an indicator of zinc status ? In : Proceedings, International Symposium on Zinc Deficiency, 29-30 April, 1982, Ankara, Turkey (in press).

Abdulla, M., Andersson, I., Asp, N.G., Berthelsen, K., Birkhed, D., Dencker, I., Johansson, C.G., Jägerstad, M., Kolar, K., Nair, B.M., Nilsson, P.E., Nordén, A., Rassner, S., Åkesson, B. & Oeckerman, P.A. (1981a) Nutrient intake and health status of vegans. Chemical analysis of diets using the duplicate portion technique. Am. J. Clin. Nutr. 34:2464-2477.

Abdulla, M., Andersson, I., Belfrage, P., Dencker, I., Jägerstad, M., Melander, A., Nordén, Å, Scherstén, B., Thulin, T. & Åkesson, B. (1979a) Assessment of food consumption. In : Nutrition and Old Age. Chemical analysis of what old people eat and their states of health during 6 years of follow-up. Scand. J. Gastroenterol. (Suppl. 52) 14:28-41.

Abdulla, M. & Haeger-Aronsen, B. (1971) Aminolevulinic acid dehydratase activation by zinc. Enzyme 12:708-710.

Abdulla, M., Jägerstad, M., Kolar, K., Nordén, A., Schütz, A., Svensson, S. (1982a) Essential and toxic inorganic elements in prepared meals - 24-hour dietary sampling employing the duplicate portion technique. In : Proceedings, Second International Workshop on Trace Element Analytical Chemistry in Medicine and Biology, 21-24 April, 1982, Neuherberg, FRG. Walter de Gruyter, Berlin & New York (in press).

Abdulla, M., Jägerstad, M., Melander, A., Nordén, A., Svensson, A, & Wahlin, E. (1979b) Iodine. In : Nutrition and Old Age. Chemical analysis of what old people eat and their states of health during 6 years of follow up. Scand. J. Gastroenterol. (Suppl. 52) 14:185-190.

Abdulla, M. Jägerstad, M., Nordén, Å, Qvist, I. & Svensson, S. (1977) Dietary intake of electrolytes and trace elements in the elderly. Nutr. Metab. (Suppl. 1) 21:41-44.

Abdulla, M., Löfberg, L., Jägerstad, M., Qvist, I., Svensson, S. & Åberg, A. (1982b) Plasma and amniotic fluid concentrations of essential chemical elements during pregnancy. In : Proceedings, Second International Workshop on Trace Element Analytical Chemistry in Medicine and Biology, 21-24 April, 1982, Neuherberg, FRG. Walter de Gruyter, Berlin & New York (in press).

Abdulla, M. & Svensson, S. (1979) Effect of oral zinc intake on delta-aminolevulinic acid dehydratase in red blood cells. Scand. J. Clin. Lab. Invest. 39:31-36.

Abdulla, M., Svensson, S. & Nordén, A. (1981b) How inadequate is the dietary intake of electrolytes and trace elements in the elderly ? In : Recent Advances in Clinical Nutrition (Ed. A.N. Howard & I.M. Baird). John Libbey, London, pp. 23-24.

Abdulla, M., Svensson, S., Nordén, A. & Oeckerman, P.A. (1981c) Dietary intake of trace elements from vegetarian and non-vegetarian diets. XII International Congress of Nutrition, 16-21 August, 1981, San Diego, USA (abstract 872).

Beisel, W.R., Pekarek, R.S. & Wannemacher, R.W., Jr. (1976) Homeostatic mechanisms affecting plasma zinc levels in acute stress. In : Trace Elements in Human Health and Disease (Ed. A.S. Prasad). Academic Press, New York, San Francisco, London. Vol. 1, pp. 87-106.

Borgström, B., Dencker, I., Krabisch, L., Nordén, A. and Åkesson, B. (1975) Assessment of food consumption. In : A Study of Food. Consumption by the duplicate portion technique in a sample of the Dalby population (Ed. B. Borgström, A. Nordén, B. Åkesson & M. Jägerstad). Scand. J. Soc. Med., Suppl. 10:14-25.

Chen, X., Yang, G., Chen, J., Chen, X., Wen, Z. & Ge, K. (1980) Studies on the relation of selenium and Keshan disease. Biol. Trace Elements Res. 2:91-107.

FAO/WHO Expert Committee on Food Additives (1972) Evaluation of certain food additives and the contaminants mercury, lead and cadmium. Report No 16, WHO Techn. Rep. Ser. No 505, pp. 1-32.

Hambidge, K.M., Hambidge, C., Jacobs, M. & Baum, D.J. (1972) Low levels of zinc in hair, anorexia, poor growth and hypogeusia in children. Pediat. Res. 6:868-874.

Hamilton, E.I. (1979) The Chemical Elements and Man : Measurements, Perspectives, Applications (Ed. K.I. Newton) C.C. Thomas, Springfield, Illinois, USA, pp. 19-25.

Hurley, L.S. & Swenerton, H. (1966) Congenital malformations resulting from zinc deficiency. Proc. Soc. Exp. Biol. Med., 123:692-696.

Jameson, S. (1976) Effects of zinc deficiency in human reproduction. Acta Med. Scand. (Suppl. 593) 197:3-89.

Jeejeebhoy, K.N., Chu, R.C., Marliss, E.B., Greenberg, G.R. & Bruce-Robertson, A. (1977) Chromium deficiency, glucose tolerance and neuropathy reversed by chromium supplementation in a patient receiving long-term total parenteral nutrition. Am. J. Clin. Nutr. 30:531-538.

Kirchgessner, M. & Roth, H.P. (1980) Biochemical changes of hormones and metalloenzymes in zinc deficiency. In : Zinc in the Environment - Part II - Health Effects (Ed. J.O. Nriagu) John Wiley & Sons, Inc., London, pp. 72-101.

Klevay, L.,M., Reck, S.J. & Barcome, D.F. (1979) Evidence of dietary copper and zinc deficiencies. J. Am. Med. Assoc. 241:1916-1918.

Mathur, A. (1978) Role of zinc in experimental and clinical oral cancer. Thesis. University of Lund, Sweden.

Mertz, W. (1970) Some aspects of nutritional trace element research. Fed. Proc. 29:1483-1488.

Mertz, W. (1981a) The essential trace elements. Science 213:1332-1338.

Mertz, W. (1981b) The scientific and practical importance of trace elements. Phil. Trans. R. Soc. Lond., B 294:9-18.

Powanda, M.C. (1982) The role of leukocyte endogenous mediator (endogenous pyrogen) in inflammation. In : Inflammatory Diseases and Copper (Ed. J.R.J. Sorenson). Humana Press, Clifton, New Jersey, pp. 31-43.

Prasad, A.S. (1976) Deficiency of zinc in man and its toxicity. In : Trace Elements in Human Health and Disease (Ed. A.S. Prasad). Academic Press, New York, San Francisco, London. Vol. 1, pp. 1-20.

RDA - Recommended Dietary Allowances : Ninth Revised Edition (1980) A Report of the Food and Nutrition Board, National Academy of Sciences, National Research Council, Washington D.C., USA.

Sandstead, H.H., Vo-Khactu, J.P. & Solomons, N. (1976) Conditioned Zinc Deficiencies. In : Trace Elements in Human Health and Disease" (Ed. A.S. Prasad). Academic Press, New York, San Francisco, London. Vol. 1, pp. 33-49.

Smith, J.C., Morris, E.R. & Ellis, R. (1982) Requirements and bioavailability of zinc. In : Proceedings, International Symposium on Zinc Deficiency, 29-30 April, 1982, Ankara, Turkey (in press).

Thulin, T., Abdulla, M., Dencker, I., Jägerstad, M., Melander, A., Nordén, Å., Scherstén, B. and Åkesson, B. (1980) Comparison of energy and nutrient intakes in women with high and low blood pressure levels. Acta Med. Scand. 208:367-373.

Underwood, E.J. (1977) Trace Elements in Human and Animal Nutrition. Academic Press, New York, 4th Edition, pp. 545.

Weissman, K. (1980) Zinc deficiency and effects of systemic zinc therapy. M.D. Thesis. FADL s förlag, Copenhagen, Arhus, Odense.

Whitehouse, R.C., Prasad, A.S., Rabbani, P.I. & Cossack, Z.T. (1982) Zinc in plasma neutrophils, lymphocytes and erythrocytes as determined by atomic absorption spectrophotometry. Clin. Chem. 28:475-480.

FOOD CONSUMPTION, NEUROTRANSMITTER SYNTHESIS, AND HUMAN BEHAVIOUR

Richard J. WURTMAN

Department of Nutrition and Food Science, Massachusetts Institute of
Technology, Cambridge, Massachusetts, U.S.A.

SUMMARY

Fluctuations in the availability to the brain of tryptophan
and tyrosine cause major changes in the rates at which neurons
synthesize serotonin and the catecholamines respectively. Such
changes occur when the pattern of neutral amino acids in the
plasma is altered by what has been eaten, the plasma changes
inducing parallel changes in the amounts of tryptophan or
tyrosine that are transported from blood to brain. Diet-induced
changes in plasma choline levels can produce similar changes in
acetylcholine synthesis.

The three nutrients, tryptophan, tyrosine and choline, when
administered in the pure form or simply ingested in food, can
thus act like drugs giving rise to important changes in the
chemical composition of structures in the brain. The interactions that relate the amount of a nutrient administered or
ingested to its level in the blood plasma, its level in the
brain and its effect on nerve neurotransmission, are not simple.
The conversion of tryptophan into serotonin is influenced by the
proportion of carbohydrate in the diet; the synthesis of serotonin in turn affects the proportion of carbohydrate an individual subsequently chooses to eat. In the case of choline and

tyrosine the effect on a neuron of an increased supply of the nutrient varies with the neuron's firing frequency and can lead to changes in that frequency. Choline and tyrosine can hence amplify neurotransmission selectively, increasing it at some synapses but not at others.

Taken together, these findings illustrate a novel and hitherto unsuspected aspect of nutrition's effects on the brain and provide the basis for new modes of therapy for patients with some metabolic, neurologic or psychiatric brain diseases.

* * *

Foods are mixtures of chemicals which are consumed electively at intervals during the day, partially metabolized within the gut, and then released into the blood stream for distribution to the tissues. Some of these chemicals, the nutrients, are continuously required by cells; at times when food is not being digested and absorbed, they are released into the blood stream from reservoirs in particular tissues. The fluxes of the major nutrients between the blood stream and most tissues are controlled by hormones released after eating. (The most important of these hormones, insulin, facilitates the passages of glucose, fatty acids, and most amino acids into the tissues; the fall in plasma insulin that occurs when foods stop being absorbed causes the net flow of these nutrients to become tissue-to-plasma). Brain is an exception; the passage of nutrients between its extracellular space and the plasma is not controlled by hormones like insulin but depends instead on the kinetic characteristics of specific transport macromolecules, located within the capillary endothelial cells comprising the blood-brain barrier, which physically carry the nutrient molecules in either direction (Pardridge, 1977). These transport systems, are, in most cases, unsaturated with their circulating ligands, hence a postprandial increase (or decrease) in the plasma concentration of a nutrient like glucose, tyrosine, or choline will facilitate (or suppress) its uptake into the brain.

Postprandial changes in plasma levels of most nutrients are relatively short-lived, since the compounds can be metabolized by the gut and liver before entering the systemic circulation (and during recirculations), or incorporated within the tissues into larger, water-soluble, "reservoir" molecules like glycogen, triglycerides, and proteins. In some cases (for example, those of dietary calcium or glucose), plasma levels of the nutrients are kept within a very narrow range postprandially, regardless of the composition of the food that was consumed, through the operation of homeostatic mechanisms in which, typically, a small change in the plasma level is "sensed", and processes are then activated which accelerate or slow the compound's removal from the plasma (i.e., metabolism or tissue uptake). In other cases (the amino acids and choline) (Hirsch et al., 1978; Fernstrom et al., 1979), no such feedback loops operate, so that plasma levels can vary across a wide range, depending solely on what is currently being digested. For example, plasma valine levels may be as much as six-fold higher after a protein-rich breakfast than after a protein-free one (Fig. 1), and plasma choline is elevated three-fold by a breakfast and lunch of eggs (Fig. 2).

My associates and I have found that some of the changes in plasma constituents (notably the amino acids and choline) that normally follow eating can have important secondary effects on the nervous system, selectively increasing or decreasing the rates at which neurons synthesize neurotransmitters from circulating nutrients (Wurtman et al., 1980; Wurtman, 1982). Moreover, these effects can sometimes be amplified by administering the nutrients (like tyrosine or choline) in pure form, as though they were drugs (Wurtman et al., 1980; Wurtman, 1982), and by mixing them with carbohydrates (Mauron & Wurtman, 1982). The brain apparently uses the ability of some of its neurons to couple neurotransmission to food-induced changes in plasma composition as a source of information about the individual's nutritional and metabolic state, and as a basis for making decisions about subsequent food intake and about such cyclic behavioural processes as sleeping. For example, if the first meal of the day is rich in carbohydrate and poor in protein, the changes that it produces in the pattern of amino acids within the plasma will raise the brain levels of a nutrient, the amino acid tryptophan, and this, in turn, will accelerate the production and release of its neurotransmitter product, serotonin (Fernstrom & Wurtman, 1971; 1972). The resulting changes in neurotransmission will cause the individual to choose away from carbohydrates and towards protein at lunchtime; a protein-rich breakfast, which diminishes brain serotonin, will have opposite effects.

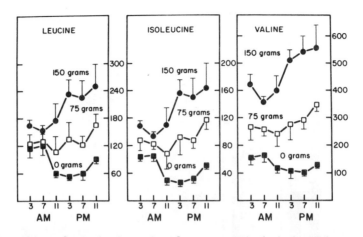

Fig. 1 : Diurnal variations in plasma branched-chain amino acid levels (nmol/ml ± SD) in 7 normal subjects ingesting different levels of dietary protein for 5 consecutive days. (Plasmas were sampled on the 4th and 5th day of each treatment period). In this and all other figures ■ = no-protein diet; □ = 75 g protein diet; ● = 150 g protein diet. Identical meals were served at 8 am, 12 noon and 5 pm (Fernstrom et al., 1979).

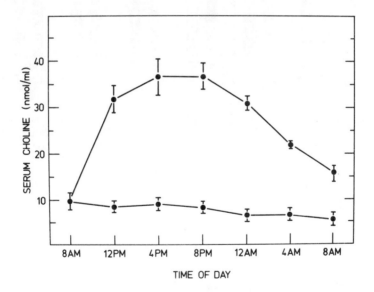

Fig. 2 : Serum choline levels during consumption of a high-choline (A) and a low-choline (B) diet by 6 normal human subjects for 2 days. Each point represents the mean ± SEM. The high-choline diet contained 50 mg/day, or about 0.67 mg/kg. Meals were eaten starting at 8.30 am, 12 noon and 5 pm. Blood samples were obtained at 8 am on day 2 and every 4 hrs thereafter to 8 am the following day. Data were analyzed by 2-way ANOVA and paired t-test; *p < 0.01 (Hirsch et al., 1978).

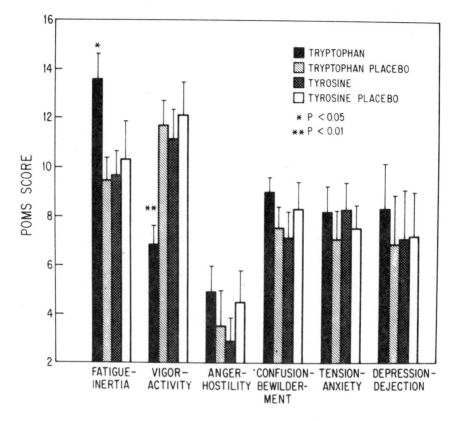

Fig. 3 : Subjects received oral tryptophan (50 mg/kg) or tyrosine (100 mg/kg), using a double-blind, placebo-controlled, crossover design. The substances were ingested at 7 am and behavioural testing began at 9 am. Mood, measured by the Profile of Mood States (POMS), was significantly altered. Specifically, tryptophan significantly decreased vigour (p < 0.01) and increased fatigue (p < 0.05) when compared to either placebo or tyrosine (Lieberman et al., 1982).

The changes in brain serotonin induced by consuming foods containing varying proportions of protein to carbohydrate can also influence other behavioural phenomena, including mood (Fig. 3), sleepiness, and pain sensitivity. Other neurotransmitters besides serotonin are also affected by nutrient availability, but the value to the individual of having, for example, production of the catecholamine neurotransmitters (dopamine, norepinephrine, epinephrine) coupled to brain tyrosine levels, or of coupling the synthesis of acetylcholine to brain choline, remains a mystery. In actuality, these latter nutrients (tyrosine and choline) affect the production of their respective neurotransmitter products only under specific circumstances. In order for particular catecholaminergic or cholinergic neurons to become responsive to levels of the precursor nutrients, the brain must cause them to fire relatively frequently. (In contrast, serotonin-releasing neurons seem always to respond to having more or less tryptophan). Just the same, the ability of tyrosine or choline to act as amplifiers, increasing the amounts of their neurotransmitter products being released from active neurons, affords these compounds considerable potential use in the treatment of diseases or conditions thought to involve catecholamines or acetylcholine.

Brain serotonin responses to dietary carbohydrate and protein

Consumption of a carbohydrate-rich, protein-free breakfast causes major changes in the pattern of amino acids in the plasma, largely because of the secretion of insulin. The pancreatic hormone facilitates the uptake of most of the amino acids into tissues like skeletal muscle, thus lowering their plasma concentrations markedly. Plasma tryptophan (Trp) levels, however, do not fall - primarily because the bulk of the tryptophan is loosely bound to circulating albumin, which retards its entry into peripheral tissues while allowing it to enter the brain (Madras et al., 1974). Hence the ratio of plasma tryptophan concentration to the plasma concentrations of other large, neutral amino acids (LNAA) like leucine, isoleucine, valine, tyrosine and phenylalanine rises markedly. This ratio largely determines the concentration of tryptophan within the brain (Fernstrom & Wurtman, 1972) because of the characteristics of the transport mechanism that carries tryptophan across the blood-brain barrier. This transport system is unsaturated with its amino acid ligand; hence an increase or decrease in plasma tryptophan will rapidly increase or decrease the amount of tryptophan being carried per unit time. Moreover, a single transport mechanism carries all of the LNAA - including tryptophan - competitively; thus an increase in the Trp/LNAA plasma ratio (caused postprandially, for example, by an insulin-mediated fall in the other LNAA) rapidly elevates tryptophan levels throughout the brain (Fernstrom & Wurtman, 1971; 1972; Pardridge, 1977).

Within those relatively few brain neurons that convert tryptophan to serotonin and use the indoleamine as their neurotransmitter, the rise in tryptophan quickly increases the substrate-saturation of the serotonin-forming enzyme tryptophan hydroxylase, thereby sequentially increasing the production of serotonin, its absolute levels within nerve terminals, and its release into synapses and the brain's extracellular space each time the neurons fire (Fernstrom & Wurtman, 1971; 1972).

If the initial meal is, instead, rich in protein, plasma tryptophan levels rise (Fernstrom et al., 1979) (Fig. 4) because some of the tryptophan molecules in the protein are able to traverse the liver and enter the blood stream. However, levels of the other LNAA not only do not fall, as after a carbohydrate meal, but actually rise manyfold (Fig. 5). This difference between tryptophan and the other LNAA reflects their relative abundance in protein. Tryptophan is scarce, comprising only about 1-1.5% of most proteins; the other LNAA, as a group, are not. It also reflects the fact that some or the other LNAA - leucine, isoleucine, and valine - are largely unmetabolized during their passage through the liver. Hence the plasma Trp/LNAA ratio falls in response to dietary protein, as do brain tryptophan and serotonin (Fernstrom & Wurtman, 1972).

Serotonin-releasing neurons can thus be conceived as "variable-ratio sensors", emitting more or less of their signal, serotonin, depending upon the chemicals that they sense (plasma concentrations of tryptophan and the other LNAA). These chemicals, in turn, vary predictably depending upon the composition of the food currently being digested and absorbed.

Serotonin neurons and the control of food choice

If animals are allowed to choose concurrently among two or more foods of differing composition, their observed behaviour suggests that appetitive mechanisms are operating which allow them to regulate not only the mass of food that they eat and its number of calories, but also the proportion of protein to carbohydrate (or carbohydrate-to-protein) within the meal. This can be demonstrated either by keeping the carbohydrate or calorie contents of the test foods constant and varying the per cent protein, or by keeping the protein and calories constant and varying the per cent carbohydrate.

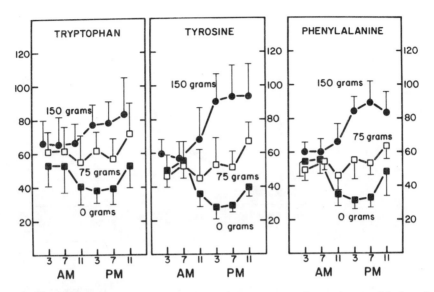

Fig. 4A : Diurnal variations in plasma aromatic amino acid levels in normal human subjects consuming different levels of dietary protein, as described in legend to Fig. 1 (Fernstrom et al., 1979).

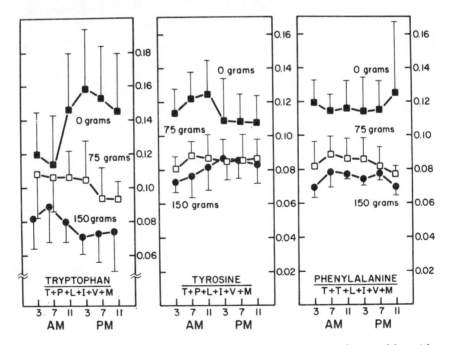

Fig. 4B : Diurnal variations in plasma aromatic amino acid ratios in normal human subjects ingesting different levels of dietary protein, as described in legend to Fig. 1 (Fernstrom et al., 1979).

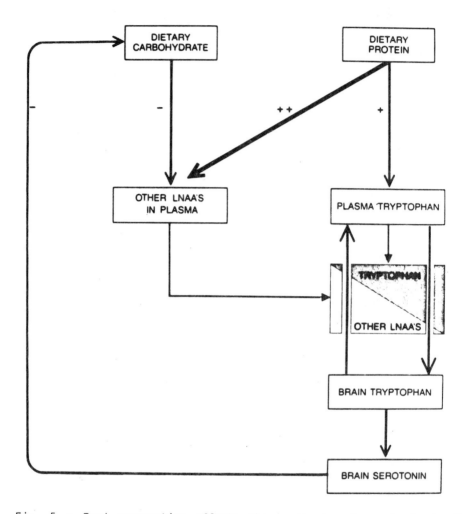

Fig. 5 : Food consumption affects the synthesis of serotonin in the brain. Eating protein raises the plasma level of tryptophan, but it raises the plasma level of five other large, neutral amino acids (LNAA) even further because each of them is more plentiful than tryptophan in proteins. Dietary carbohydrate induces the secretion of insulin, which moves most amino acids out of the blood stream but has little effect on tryptophan. Because tryptophan must compete with the other LNAA for transport across the blood-brain barrier, the movement of tryptophan from the plasma to the brain is controlled by the plasma ratio of tryptophan to the other LNAA. When the ratio is high, tryptophan enters the brain; when it is low, tryptophan moves from the brain into the blood stream. The release in the brain of serotonin synthesized from the tryptophan appears to reduce a rat's (or a person's) carbohydrate intake.

.(That is, one can demonstrate regulation of both the protein and the carbohydrate components of the diet). The ability to choose among foods so as to obtain a "desired" per cent of carbohydrates (per cent of total mass or per cent of calories) is apparently independent of the sweetness of the carbohydrate being tested. Regulation is as easy to demonstrate for dextrin as for dextrose or sucrose (Wurtman & Wurtman, 1979).

Serotonin-releasing neurons apparently are key components of the brain mechanisms underlying this regulation of food choice. Thus, rats can be caused to choose away from carbohydrate (i.e., to increase the protein/ carbohydrate ratio of the test meal) either by giving them a carbohydrate-rich "pre-meal" (by stomach tube) (Wurtman et al., 1983) or by administering any of a large number of drugs which act at different loci to enhance serotonin-mediated neurotransmission (Wurtman & Wurtman, 1979). Moreover, dietary manipulations (like the chronic consumption of a carbohydrate-poor diet) which diminish the plasma Trp/LNAA ratio, and thus reduce brain tryptophan and serotonin levels, sharply increase the proportion of carbohydrate that the animal subsequently eats when it is given the opportunity to choose (Wurtman et al., 1983). It seems not unlikely that the tendency of humans to consume a more-or-less constant proportion of their calories as protein, and the relative stability of lean body mass, may reflect the operation of this mechanism; it diminishes the likelihood that meals imbalanced in one direction (too much protein or carbohydrate) will be immediately followed by others imbalanced in the same direction.

It also seems possible that some of the common abnormalities in eating behaviour - including those that can lead to obesity - may have their origin in disturbances in the mechanism through which the brain "learns" about the composition of the last meal (by nutrient-induced variations in serotoninergic neurotransmission) and "decides" what to eat next. Such disturbances could exist at several loci, for example the plasma Trp/LNAA ratio might not respond appropriately to food-induced changes in the plasma Trp/LNAA ratio because of an abnormality in the blood-brain barrier transport system; the increase in brain tryptophan that follows a carbohydrate-rich, protein-poor meal might not have a sufficient effect on serotonin-mediated neurotransmission because the unfortunate individual has a brain that contains too few

serotoninergic neurons : the functional activity of the serotoninergic
neurons may be disturbed by pathological processes arising elsewhere. While
it is difficult to examine changes in brain chemistry using human subjects,
it is relatively easy to determine whether, for example, obese individuals
exhibit the same changes in the plasma Trp/LNAA ratio after a test meal as
non-obese control subjects, and we are currently exploring such responses in
collaboration with associates as the University of Lausanne. We have during
the past few years identified a subset of obese people whose problem seems
to be specifically related to an inappropriate desire for carbohydrates,
especially at certain times of day (Wurtman et al., 1981). If such "carbo-
hydrate-cravers" compensate by diminishing their mealtime food intake or by
increasing their energy output they may avoid the development of obesity;
however, if they eat normal-sized meals supplemented with multiple carbohydrate snacks they become obese. We find that administration of sub-
anorectic doses of drugs (like fenfluramine) which release brain serotonin
can suppress the carbohydrate craving (Wurtman et al., 1981). It will be
interesting to see whether supplemental tryptophan has a similar effect,
especially in view of the evidence that the Trp/LNAA plasma ratio tends to
be abnormally low in obese people, possibly as a consequence of peripheral
insulin resistance, and that the protein-rich, carbohydrate-poor diet (the
"PSMF diet") with which many obese people are treated further lowers the
Trp/LNAA ratio (Heraief et al., 1983).

Other nutrient-dependent neurotransmitters

As mentioned above, the rates at which catecholaminergic neurons release
their transmitter (dopamine, noradrenaline or adrenaline) can be enhanced,
if the neurons are firing frequently, by giving the individual tyrosine, and
can be suppressed by administering other LNAA (which compete with tyrosine
for uptake into the brain). The effect of a given tyrosine dose (or, for
that matter, of a given dose of tryptophan) can be enhanced by giving it
concurrently with a carbohydrate source (Mauron & Wurtman, 1982). The
resulting secretion of insulin lowers plasma levels of the competing LNAA,
and increases the proportion of administered molecules that are taken up
into the brain (Pardridge, 1977; Mauron & Wurtman, 1982). The reason that
tyrosine levels affect catecholamine synthesis only when neurons are firing

rapidly has to do with the properties of the key enzyme, tyrosine hydroxylase, that converts the amino acid to the catecholamines (Wurtman et al.,
1980; Wurtman, 1982). When neurons fire frequently, the enzyme itself
becomes phosphorylated; as a consequence its affinity for its cofactor,
tetrahydrobiopterin, increases markedly, so that it becomes limited by
tyrosine. This phosphorylation is short-lived, hence soon after the neurons
slow their firing they become unresponsive to additional tyrosine. Tyrosine
administration can, theoretically, be useful in any experimental or clinical
situation in which it would be desirable to have more catecholamine molecules released at a particular locus, inside or outside the brain. (Its uses
in these situations, and the theoretical bases for such uses, have been
reviewed extensively (Wurtman et al., 1980).

Similarly, acetylcholine's production in and release from cholinergic
neurons can depend on the availability of free choline, when particular
cholinergic neurons happen to be firing frequently. Any treatment that
increases plasma choline, including oral choline chloride, dietary lecithin,
or even eating eggs or liver, will have this effect. The primary reason that
brain choline levels rise after choline consumption is not that more plasma
choline enters the brain but that less leaves it. The brain is able to
synthesize choline de novo (Blusztajn & Wurtman, 1981; 1983); however, most
of the choline is lost by secretion into the blood stream via the operation
of a bi-directional blood-brain barrier choline transport system. Raising
plasma choline slows this loss, thereby allowing brain choline to
accumulate. Choline is converted to acetylcholine by the enzyme choline
acetyl-transferase; the mechanism that couples the choline-dependence of
this process to neuronal firing frequency awaits identification. Choline (or
lecithin) administration has been found to be useful in treating tardive
dyskinesia, mania, and several neurological diseases (Wurtman et al., 1980).
It may also constitute a useful adjunct in the management of Alzheimer's
disease, or senility - if and when an effective drug is found that the
choline can be an adjunct to, - since the primary lesion in this disease
appears to be a selective loss of acetylcholine-releasing brain neurons.

Although the effects of tyrosine and choline on brain function are less
ubiquitous than those of tryptophan, when they do occur (i.e., in rapidly-

firing neurons) they can be of major health importance, for example, in memory, or in control of mood, or in sustaining normal cardiovascular function. Hence it seems prudent that care should be given to satisfying the brain's needs for these compounds - especially in formulating foods for people who might already have some disturbance involving catecholaminergic or cholinergic neurons (like the aged). The fact that tyrosine may not be an essential amino acid, nor choline an essential growth factor, for the young rat should not obscure our recognition that these compounds, like trypto- phan, are absolutely essential for a normally-functioning nervous system.

ACKNOWLEDGEMENTS

These studies were supported in part by grants from the United States National Institutes of Health, the National Aeronautics and Space Adminis- tration, and the Center for Brain Sciences and Metabolism Charitable Trust.

REFERENCES

Blusztajn, J.K. & Wurtman, R.J. (1981) Choline biosynthesis by a preparation enriched in synaptosomes from rat brain. Nature 290:417-418.

Blusztajn, J.K. & Wurtman, R.J. (1983) Choline and cholinergic neurons. Science (in press).

Fernstrom, J.D. & Wurtman, R.J. (1971) Brain serotonin content : Increase following ingestion of carbohydrate diet. Science 174:1023-1025.

Fernstrom, J.D. & Wurtman, R.J. (1972) Brain serotonin content : Physio- logical regulation by plasma neutral amino acids. Science 178:414-416.

Fernstrom, J.D., Wurtman, R.J., Hammarström-Wiklund, B., Rand, W.M., Munro, H.N. & Davidson, C.S. (1979) Diurnal variations in plasma concentrations of tryptophan, tyrosine and other neutral amino acids : Effect of dietary protein intake. Am. J. Clin. Nutr., 32:1912-1922.

Heraief, E., Burckhardt, P., Mauron, C., Wurtman, J.J. & Wurtman, R.J. (1983) Obesity and its dietary treatment may suppress synthesis of brain serotonin. Submitted to J. Clin. Endocrinol. Metab.

Hirsch, M.J., Growdon, J.H. & Wurtman, R.J. (1978) Relations between dietary choline intake, serum choline levels, and various metabolic indices. Metabolism 27:953-960.

Lieberman, H.R., Corkin, S., Spring, B.J., Growdon, J.H. & Wurtman, R.J. (1982) Mood and sensorimotor performance alter neurotransmitter precursor administration. Society for Neuroscience, 8, 395 (abstract).

Madras, B.K., Cohen, E.L., Messing, R., Munro, H.N. & Wurtman, R.J. (1974) Relevance of serum-free tryptophan to tissue tryptophan concentrations. Metabolism 23:1107-1116.

Mauron, C. & Wurtman, R.J. (1982) Co-administering tyrosine with glucose potentiates its effect on brain tyrosine levels. J. Neural Transm. 55:317-321.

Pardridge, W.M. (1977) Regulation of amino acid availability to the brain. In : Nutrition and the Brain (Ed. R.J. Wurtman & J.J. Wurtman) Raven Press, New York, volume 1, pp. 141-204.

Wurtman, R.J. (1982) Nutrients that modify brain function. Scientific American 246:42-51.

Wurtman, J.J. & Wurtman, R.J. (1979) Drugs that enhance central serotoninergic transmission diminish elective carbohydrate consumption by rats. Life Sci., 24:895-904.

Wurtman, R.J., Hefti, F. & Melamed, R. (1980) Precursor control of neurotransmitter synthesis. Pharmacol. Rev. 32:315-335.

Wurtman, J.J., Wurtman, R.J., Growdon, J.H., Henry, P., Lipscomb, A. & Zeisel, S.H. (1981) Carbohydrate craving in obese people : Suppression by treatments affecting serotoninergic transmission. Int. J. Eating Disorders 1:2-15.

Wurtman, J.J., Moses, P.L. & Wurtman, R.J. (1983) Prior carbohydrate consumption affects the amount of carbohydrate that rats choose to eat. J. Nutr. 113:70-78.

GENERAL REMARKS
Some personal reflections

Arnold E. BENDER

Nutrition Department, Queen Elizabeth College,
University of London, London, England

INTRODUCTION

This symposium, dealing with nutritional requirements, the availability of nutrients in foods and the problem of their adequacy and excess, covers a broad area of the whole field of nutrition. There are many aspects that can be discussed in great depth, but my task in these general remarks is to select only some of them.

Fig. 1 illustrates the breadth of the subject and our interests. Nutritionists are interested in the entire field of food, starting even before food production, where we share our interests obviously with people like geneticists, into production, whether of animals or crops or single cell proteins, right through the properties of food, their processing and their preparation, mainly the domain of food scientists and technologists, until we reach the part which is really the basis of our discussion, the nub of the main nutritional field - "the nutrient requirements". The term "food selection" covers a vast amount of human knowledge involved in answering the question : "Why do we eat what we do ?". Rather than discussing each presentation separately, I shall present my views on some of the problems involved in determining nutrient requirements.

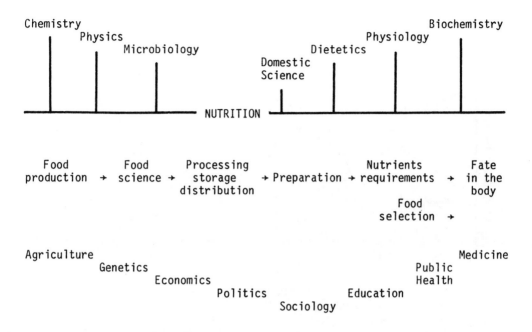

Fig. 1 : Scope of nutrition

An optimum diet

One of the questions requiring an answer is : "How much food do we need for what purpose ?". Fig. 2 describes the different criteria that can be used to establish recommended dietary allowances (RDA). Thus, we can recommend an amount of a nutrient that prevents the appearance of clinical signs of deficiency, or the greater amount needed to saturate an appropriate enzyme system or to saturate the tissues.

However, between adequate nutrition and sub-clinical malnutrition is a stage which I have termed "covert" or "hidden" malnutrition, which only comes to light under stress, and the obvious example is ascorbic acid (vitamin C) : 10 mg a day will cure scurvy, and 20 mg are required for adequate wound healing. Investigations in Bangladesh some three years ago, showed that an increase in hospital admissions for xerophthalmia followed the rice harvests. There was a major peak after the main rice harvest and a smaller peak after the second one; the explanation suggested was that when children were so poorly fed that they did not grow, they showed no signs of

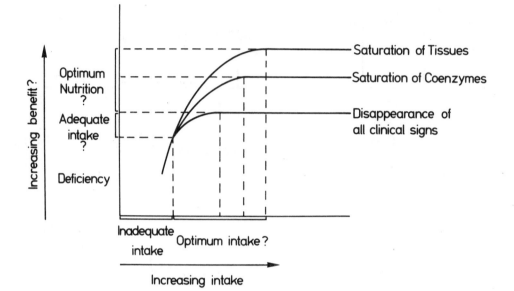

Fig. 2 : Minimum, adequate and optimum nutrient intake.

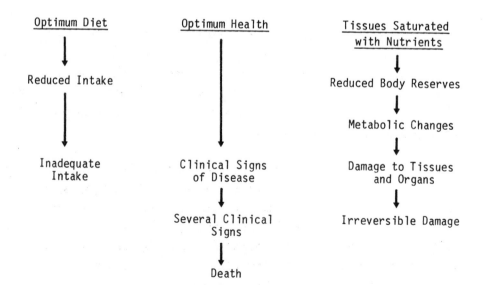

Fig. 3 : Nutrient intake and nutritional status.

xerophthalmia, but when they were better fed and able to grow, then they developed xerophthalmia.

Three methods exist whereby one can measure nutritional status : dietary intake (not an index of status but merely confirmatory), clinical signs and biochemical change (Fig. 3). By definition, the optimum diet, whatever it be, would permit optimum health and the tissues would then probably be saturated with nutrients. A reduction in nutrient intake would result in a fall in body reserves, but we have no evidence on the optimum levels of body reserves. Liver stores of vitamin A must be reduced almost to exhaustion before blood levels fall. Leucocyte levels of vitamin C can vary enormously without detectable adverse effects, as can tissue levels of all nutrients. Consequently it is not an easy matter to establish nutritional adequacy of the diet. Dietary experiments inevitably suffer from the problem that any change in the level of a major ingredient must be matched by a corresponding change in another. So, experimentally it is not possible to change only one factor at a time.

Several other experimental difficulties exist when trying to define nutrient requirements. Firstly, we have the problems of individual variation and fluctuation. An example is that shown by W.C. Rose when he established the amino acid requirements. In a group of 30 subjects the amount of lysine needed to maintain nitrogen balance differed by 100% between the lowest and highest responders. Another problem is adaptation. For many years nutritionists disbelieved figures coming from places like Papua - New Guinea, of women successfully lactating on 1,800 kcal. This has now been confirmed and Dr. Whitehead's own similar figures have been observed by others too. So, there seems to exist a degree of adaptation which enables people to live on diets with nutrient levels very much below our recommendations.

Individual variation

Some 10 years ago Durnin and Southgate, in re-examining energy conversion factors in human beings, found a considerable variation in the individual's ability to digest cellulose and hemi-cellulose (although we call these indigestible matter). Dr. Widdowson mentioned that subjects differed two-fold in energy intakes. Fig. 4 shows an even greater range. As regards energy, we assume that if the subjects are maintaining their weight then they must differ up to four-fold in their energy needs. If all were consuming a similar diet, then subjects at the lower end of the scale must be ingesting only one-quarter of the protein, minerals and vitamins of those at the upper end of the scale. We do not know if their requirements differ to this extent.

Indeed, some of our problems may arise from our frequent attempts to seek a single answer in situations as above. There are nutritionists who lay most if not all the blame on a single nutrient or lack of it whether it is vitamin C, saturated fat, sugar or dietary fibre. It is more likely that different people respond differently and to a different extent to dietary change and to the environment. For example, we have been discussing whether man has functional brown adipose tissue but it is possible that some people have and that others have not. In other words, there might be many different types of people, many different types of answers and when Dr. Widdowson asks "why do some people eat twice as much as others", one can offer several potential explanations, all of which or any one of which might be true. Thus, some people certainly eat to the capacity of their stomach and, in the middle of a meal, will cease eating. Others seem to have an almost infinite stomach size. Then there are some, usually obese people, who find a delight in swallowing and will eat whatever is available because of the pleasure of swallowing. Behaviourists tell us of people who eat with their eyes; if they see it, they eat it and will not leave the table if there still is food there. Then there are some with abnormally low metabolic rates, and we know of people who actually gain weight on 1,500 kcal. a day. A possible but only partial answer to some of these situations is a change in metabolic rate during sleep.

According to Dr. Jéquier, there is a larger loss of energy from a faster infusion of glucose; this would imply that speed of eating might have an effect on some people. Certainly, meal frequency does and we heard about insulin resistance, caffeine and smoking. Then obviously, people differ in their exercise capacity and efficiency of movement and hence energy expenditure.

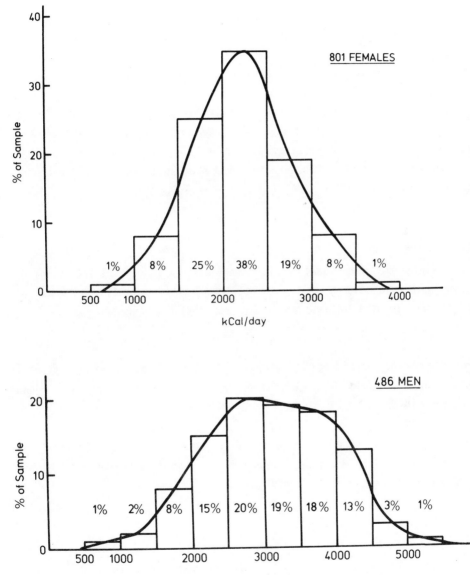

<u>Fig. 4</u> : Range of calorie intake (compiled from Harries et al., Proc. Nutr. Soc. 1962, 21:157-169).

Dr. Widdowson's figures showed that young children's energy intake has not fallen over the years, whilst that of older children has. I wondered whether breakfast might be the answer. We have observed that children aged 5 all eat breakfast; by the time they are 15, 25% omit breakfast. We have no older figures for breakfast consumption; there is no way of comparing numbers missing breakfasts today with a generation ago. We carried out an age-effect survey on a thousand people, and found a straight line relationship. At the age of 20, about 25% had no breakfast but this figure fell with increasing age up to 65, when nearly all had breakfast. Whether that means as you get older you feel more like eating breakfast, or whether in the old days one always had breakfast and hence still continues to do so, we do not know.

Individual fluctuation

In addition to individual variation there is individual fluctuation. Although in the laboratory we try to obtain a steady base line in measurements on experimental animals and human subjects, this condition does not appear to hold in normal life.

Dr. D.A. Bender measured a number of blood parameters at fortnightly intervals over 6 months in a control group of healthy young men and found fluctuations in the blood cholesterol levels of ± 15%.

We rarely measure a base-line for so long a period and this figure illustrates how we could be misled if dietary changes or drug treatment coincided with a peak or trough in such fluctuations.

Adaptation

The problem of determining needs is further complicated by adaptation. We found that adult rats could maintain their nitrogen balance on 5% casein, and when their food intake was reduced, each individually to 70% of what they had been eating, they went into negative nitrogen balance but for only 17 days. By then they had returned to nitrogen equilibrium on the now low-protein and low-energy intake, and had lost only 3 grams in weight. They had adapted to their new food situation by reducing their oxygen consumption and so presumably, their energy expenditure by 30%.

Whether all adaptations are advantageous is not clear, particularly with reference to energy intake and diet-induced thermogenesis. It is suggested that in primitive man it was an advantage to burn off the surplus food so as to remain lean for flight and fight. On the other hand, it was a disadvantage because he was wasting food which was mostly in short supply. The interesting point is that we, in our industrialized communities, have changed everything in the last few years. Before the turn of the century, there was a famine somewhere in the world every other year. In Europe, there was a famine one year in ten and it is only in recent years that food technology has made food of all types available, from all parts of the world, all the time. So we have changed the situation only in the last one or two generations and at the same time in the Western world, it is fashionable to be slim and unfashionable to be even slightly overweight.

Once again on the question of energy - a point that remains unanswered is the effect of exercise on resting metabolism. In 1935 measurements were made on a Harvard Football Team in action and the results indicated that some 3 days later their resting metabolism was still elevated, from which one would conclude that the ordinary measurements were not correct because there is a lag later on. Those working in the energy field tell me that the how and why of this has still not been resolved.

Estimation of dietary intakes

Estimation of dietary intake using food tables, can only be taken as approximate. Different food tables differ in the values given for seemingly the same foodstuff and, perhaps more importantly, the foods for which the composition is given in the tables are rarely exactly the same as those eaten. More accuracy can be obtained by analysing the diets consumed. In addition, another source of error in the calorie evaluation are the different methods used to determine dietary fibre and the slightly different calorie factors (Widdowson in McCance & Widdowson's "The Composition of Foods" by Paul & Southgate, 1978). A threefold difference occurs between the energy contents of foods such as fruits and vegetables between UK and USA tables because of the different methods of calculation. Fortunately, the differences between the values of the major energy foods are small. It has recently come to light that the German and the United Kingdom conversion factors for fat are 38 and 37kJ, respectively (a 3% error); for carbo-hydrates, the factors are 17 and 16 kJ (as 3% error) and for alcohol, it is 30 and 29, a 3% error.

Nutrient interactions

Under controlled conditions we attempt to feed a relatively constant diet. Any change, including increased levels of nutrient(s), has repercus-sions. Increased dietary fibre, for example, may reduce the availability of calcium, iron and zinc. Increased intake of polyunsaturated fatty acids calls for an increase in vitamin E. Some nutrients can have toxic effects when ingested in excess; our normally widely fluctuating nutrient intake might enable us to avoid such repercussions. In the midst of several variables influencing intake, requirements, exercise and expenditure, and each one in turn being influenced by environmental factors, it is clearly ever more difficult to decide whether the diet of any group of individuals is adequate.

Bioavailability

Measuring the amount of a nutrient in the food chemically does not reveal whether it is bioavailable. For some nutrients we have figures from which we can make a conversion. Retinol, for example, is measured in retinol equiva-lents; thus, ß-carotene is 1/6th retinol-equivalent, and α-carotene and cryptoxanthin are 1/12th retinol-equivalent. Such estimations do not reveal how much we are going to absorb; for example, on a low-fat diet less is absorbed, on a low-protein diet it is not transported, and with no vitamin E (at least in certain experimental animals) it is oxidized in the intestine. Variations in the bioavailability of a nutrient are, therefore, very considerable but generally not taken into account.

The true assay for bioavailability is the biological one; unfortunately all bio-assays are subject to enormously wide variation in results. Physical-chemical tests are precise in the sense of being reproducible and easy to carry out and therefore quick and cheap. On the other hand, they do not necessarily measure the true nutrient content, rather an index of it. When we have verified and standardized chemical measurements against biological assays, we tend to accept them; but even then, we are most likely talking about the availability to the rat. We still know little about the bioavailability of the nutrient in question to man and that is why it is so important that Dr. Hallberg is measuring bioavailability (of iron) to human subjects.

The availability of dietary iron has always been recognised as a difficult problem more so than with other nutrients. In one experiment on soya, Layrisse found a range of absorption from 2% to 42.2% in 28 subjects. These differences can possibly be explained by the different iron statuses of the individuals. To overcome these problems in human studies absorption values are now corrected according to the absorption of a standard dose of ferrous ascorbate.

Bioavailability may become a consideration in the future with novel foods. A novel food is defined as one that we have not hitherto eaten at all or only in small quantities, and irregularly at that. If we introduce in our diet a food already eaten elsewhere, it is still novel for us. In these cases we would not be satisfied with a chemical test for the true biological content of the nutrient if eventually the food was to become a significant part of our diet.

Food processing and nutrition

Drs Hurrell and Finot discussed the changes to protein quality, loss of available lysine and the possible development of "toxins" on heating and other processing of foods. We still have many unknowns in this area.

As they pointed out, protein quality and nutritional value are important in baby foods as babies often rely on a single manufactured product as their sole source of nourishment. For adults, eating a variety of foods, nutritional quality is far less important.

People buy food because they like it; therefore the manufacturer produces foods of the right flavour and texture. Nutrition is still of very little interest to most people and even many of us do not select food on its nutritional basis; we select the food we like. And yet, to our knowledge nobody is suffering from eating processed foods or even traditionally cooked foods and any nutritional damage therefore, is probably unimportant. One might say that changes on processing are investigated because nutrition labelling is already with us, even if only on a voluntary basis. So people are becoming interested in nutrition and are demanding such information from the manufacturer. The food industry is hence becoming more involved in nutrition and also because of new information linking diet in some way with modern diseases.

Looking back at processes now in operation such as canning, drying and freezing, we must conclude that even where nutrients are partly destroyed or inactivated, such losses cannot have any major effects on nutritional health. Looking forward, however, to novel foods and novel processes such as extrusion cooking and irradiation we need to know whether they are likely to alter or affect nutritional status. They may be important if the novel foods replace traditional foods in quantity or if the novel processes are applied to traditional foods.

The consumer is also asking questions about the safety of foods. This is an area of great difficulty because many foods, indeed all cooked foods, contain identifiable toxins even if only in very small amounts. We already have different standards for processed as distinct from so-called natural foods. Natural substances extracted from a food and subsequently used as processing aids or additives are subject to close investigation and strict regulations. It is not possible to apply such conditions to unprocessed foods. Hitherto we have been engaged in examining the safety of additives, food colours, etc., all of which can be fed to experimental animals at very high dose levels, we are now faced with problems of the safety of the foods themselves which cannot be fed at high levels.

Nutritional adequacy

Is it safe was my last question, and the original question was "What do we need for optimum health, are we being adequately fed ?". I suppose the short answer is : "Since people tend to be living somewhat longer we cannot be doing all that badly !"

With the problems of determining adequacy of the diet in respect of all essential nutrients, it is becoming vital to seek information from measurements "in the field" as described by Dr. Jean Mauron. For example, vitamin requirements are determined in healthy subjects without deficiency signs and living on a good diet; but requirements must be completely different in people in a slightly less healthy state or those who have adapted to different (lower) intake levels. In real life the total diet may be poor, the environment suspect and dietary deficiencies chronic and not acute. Observations under such conditions may better serve to answer questions of adequacy than laboratory experimentation.

CONCLUSION

This symposium has given us up-to-date summaries of the particular areas of human nutrition concerned with "Nutritional Adequacy, Nutrient Availability and Needs". In common with most scientific meetings of this kind, it has generated many questions. Each of these questions now deserves further scientific research and maybe also requires a full symposium to itself, not in the expectation of answers but in the hope of directing the research and observations into useful and profitable channels. One can list some, only some, of the areas that we would want to discuss further :

- Energy intake versus needs;

- Assessment of nutritional status;

- Protein turnover in normal and abnormal conditions;

- Adaptation to low intakes of energy and other nutrients;

- Food selection and nutritional results;

- Safety assessment of foods;

- Bioavailability of nutrients;

- Role and relative importance of nutritional knowledge in influencing consumer selection;

- Optimum growth rates, disease and longevity.

Ladies and Gentlemen, we can agree that this has been a successful symposium and can thank the Organizing Committee for their skill in putting together this meeting.